THE
CONSEQUENTIAL
FRONTIER

THE
CONSEQUENTIAL
FRONTIER

CHALLENGING THE
PRIVATIZATION OF SPACE

PETER WARD

MELVILLE HOUSE
Brooklyn | London

THE CONSEQUENTIAL FRONTIER

First published in 2019 by Melville House Publishing
Copyright © Peter Ward, 2019
All rights reserved

First Melville House Printing: October 2019

Melville House Publishing
46 John Street
Brooklyn, NY 11201
and
Melville House UK
Suite 2000
16/18 Woodford Road
London E7 0HA

mhpbooks.com
@melvillehouse

ISBN: 978-1-61219-800-2
ISBN: 978-1-61219-801-9 (ebook)

A catalog record for this book is available from the Library of Congress
Library of Congress Control Number: 2019945808

Designed by Betty Lew

Printed in the United States of America

1 3 5 7 9 10 8 6 4 2

To Seren, my star,
and Leeloo,
my tiny assistant.

Contents

THE
CONSEQUENTIAL
FRONTIER

Introduction

FOR THOUSANDS OF YEARS, SPACE BELONGED TO THE GODS. Their names still loom large today—each planet in our solar system but our own is named for a Roman or Greek deity, and we continue to invoke ancient myths and legends through our spacecraft, from China's Jade Rabbit Moon lander to the Ares rocket. But just as technology and scientific inquiry eradicates mythology and superstition here on Earth, the development of space technology like telescopes, rockets, and probes have increased our understanding of the cosmos greatly, and the gods have lost some of their power. Space is now the domain of humanity.

We first stepped out into the cosmos as explorers. We marched into the cold abyss of space in the name of science, as nations flung their best and bravest off our planet to find the truths of our universe. But the mission is changing rapidly. While exploration efforts continue further afield, the tiny band of space close to our home is fast becoming a place of business, profits, and private companies. Corporations could soon rule the cosmos, and for all the dramatic warnings in science fiction, nobody is paying enough attention to the consequences.

* * *

WHEN NEIL ARMSTRONG AND BUZZ ALDRIN FIRST PLANTED their boots on the surface of the Moon in 1969, humanity collectively paused to wonder at the achievement. An estimated 600 million people around the world watched the black-and-white broadcast of man's first steps on a celestial body, the high point of a 76-hour mission there and back.[1] The Moon landing opened up a host of possibilities in space, and many imagined the inevitable Moon bases, settlements on Mars, and manned exploration missions to the further reaches of the solar system that would surely follow. But over 50 years later they are still waiting. Extraordinary feats have been accomplished since—we've landed on a meteor, sent a probe out of the solar system, and put robots on Mars—but nothing has captured the world's amazement like the Moon landing, and the startling pace at which America reached the lunar surface.

The race to the Moon was fueled by the Cold War—two world superpowers fiercely competing to reach the cosmos first and establish dominance in space. When relations between the United States and the Soviet Union thawed, the impetus was removed and progress stalled. Now that thrust has returned in another form. The commercial space sector has emerged, bringing with it reusable rockets, lower prices, and a different reason to explore space—profit. But when driven by money, mankind has historically made decisions that in hindsight reveal themselves to have been shortsighted and detrimental to the species. In space, the consequences of such decisions would be just as disastrous out there as here on Earth.

When we send humanity off this planet and out into the solar system on anything approaching a permanent basis, there will be two key questions: What do we take with us, and what do we leave behind? Spreading out into the universe allows the species to examine itself closely, to look at its faults, its mistakes, and its petty failings, and decide to start over without them. It's hard to do that when the primary motivation for those taking us to the stars is making money back here on Earth.

Without careful thought, monopolies will be built, profit margins widened, and inequality spread. Space is a tricky topic to consider (and write about)—the industry is moving extremely fast toward goals that are a long way away. Focus too closely on the short term and you'll find yourself behind the times, and look too far into the future, and you'll be dismissed as another star-gazing science-fiction fan. The fact is that major changes are happening now, spurred on by the demands of industry and economics. And so we must decide if we want to hand over the fate of space to private companies, as we have done so problematically here on Earth. Money is driving this race. It's fueling the rocket launches, funding the startups, influencing the right politicians, and buying access to the cosmos. With so much money involved, we should all hope that science, exploration, and the good of the species isn't lost on the journey, or the effects back here on Earth will only be negative.

One of the principal arguments against going to space is that we should solve our earthbound problems first. It's a simplification of a much more complex issue, and one I don't subscribe to. There will never be a perfect time to move further into space, no utopian future on Earth where the only possible next step is to go beyond our terrestrial home. Space shouldn't be viewed as a luxury to be undertaken after every other achievement is accomplished, and we must explore it for its potential to help with these problems. Nothing brings the species together with a sense of pride and victory like a major milestone in space—and anything accomplished off the planet, when done in the name of humanity rather than private interests, is a success for every person on the planet.

Many of the technologies being fired off into space already stand to benefit us here on Earth. The new wave of microsatellites can bring connectivity to people all over the world, a major factor in reducing inequality. Health breakthroughs in space can help millions, and even technology designed to keep us alive on other planets will come in handy as climate change continues to ravage ours. I hold a lot of hope for the future of humanity in space, and applaud wholeheartedly the efforts to become a multi-planetary species. But the potential to do amazing things for people here on Earth only makes it more important to get it right—and the commercial space industry is taking off so fast, it may be hard to catch up.

Just like a rocket blasting off the face of the planet, the commercialization of space has taken place in stages. The heavy lifting that got humans off the ground initially was done by NASA and the governmental agencies of Russia and Europe. NASA has probably done more than any other agency in the world to advance the achievements of humanity in space, but has struggled with a lack of funding, the dwindling attention of the American public, and bureaucracy. The Apollo program, which took Armstrong, Aldrin, and 10 others to the surface of the Moon, brought great success right up until it was ended in 1975, as the Russian and American space agencies began working together.[2] It's successor, the Space Shuttle program was approved by President Richard Nixon in 1972, and first launched in 1981. The Shuttle program's main use turned out to be helping build the International Space Station (ISS), one of the greatest feats of global cooperation, which still orbits the planet today. Despite a long list of accomplishments, including becoming the first-ever reusable spacecraft and launching many satellites and telescopes into space, the Shuttle period of NASA disappointed because it failed to build on the successes of Apollo. It was also beset with problems, which twice proved to be fatal. The Shuttle program, with no real alternative, continued beyond its expected retirement date and flew its last mission in 2011.[3]

At that point NASA found itself in an odd position—for the first time since the 1960s, it had no way of reaching space. The next generation of NASA spacecraft was far off, and the agency was forced to do two things: one was to hitch a ride on Russian rockets headed to the ISS, the other was to seriously consider commercial launch partners to send resupplies, satellites, and astronauts into space. The door for commercial companies looking to establish themselves in the launch industry was now wide open, and in walked a couple of billionaires with different philosophies on space.

Once the first stage of a rocket has spent its fuel and delivered its payload to a certain height, it falls away and the second and later stages take over. When NASA lost its ride into space, Elon Musk and Jeff Bezos had already made significant progress on their space startups, SpaceX and Blue Origin, and were eager to take over. The two entrepreneurs had made their fortunes elsewhere, but both carried a lifelong fascination with space and, over the years, developed vastly different ideas on how it can benefit mankind.

Musk has often spoken of his desire to retire on Mars and has made it his life's goal. He heartily endorses the Plan B point of view, which was perfectly summed up by Carl Sagan when he said all civilizations either become spacefaring or extinct. Musk continues to be motivated by the fear of humanity here on Earth imploding or causing another dark age. "It's important to get a self-sustaining base on Mars because it's far enough away from Earth that it's more likely to survive than a moon base," Musk said onstage at the 2018 SXSW festival.[4] "If there's a third world war we want to make sure there's enough of a seed of human civilization somewhere else to bring it back and shorten the length of the dark ages."

Bezos, on the other hand, believes humanity needs to go into space to save Earth, not humanity. "I hate the Plan B argument," he said.[5] "I think Plan B with respect to Earth being destroyed is make sure that Plan A works. We've sent robotic probes to every planet in this solar system, believe me, this is the best one. We know that. It's not even close. My friends who say they want to move to Mars or something . . . I say, 'Why don't you live in Antarctica for a year first, because it's a garden paradise compared to Mars."

SpaceX and Blue Origin are not the only rocket companies in the game, but they attract the most attention due to their achievements and high-profile founders. This difference in philosophy has expressed itself in differing approaches to space. While Musk has made outlandish predictions, called for drastic measures, and celebrated each success with boyish enthusiasm, Bezos's Blue Origin has been going about its task much quieter, preferring to make steady progress under the radar rather than flashy, live-streamed launches and ambitious timelines. In May 2019, Bezos bucked that trend slightly, when he unveiled the Blue Origin moon lander, and announced he wanted to move all of the Earth's heavy industry and mining into space, to preserve our planet and fight off climate change. This idea would see Earth become "zoned" for light industry and residential purposes only. He also revealed his ambitions to launch a constellation of gigantic space stations where people would live, alleviating overpopulation.

In some ways these companies are extensions of their founders' characters, with Musk the eccentric adventurer and Bezos the more calculating industrialist. If both were to carry out plans in space similar to what they've already achieved on Earth, Musk could recklessly strike out for Mars first, while Bezos would meticulously gain a stranglehold over crucial infrastructure closer to home. Neither situation is ideal, but if we're going to make real progress in the solar system, it seems inevitable these two men will play a significant role.

* * *

TALK OF STORES, FACTORIES, AND TOWNS ON OTHER PLANETS raises comparisons to the emergence of economic hot spots here on Earth. Perhaps the most pertinent analogy, and certainly the most used, is the spread of railroads across America in the late nineteenth century. Railroad expansion was fairly slow until private companies were given the means and the resources to build links across the country and make huge profits in the process. The expansion was hailed as a great success for the private sector, pushing through a frontier with the full force of capitalism, and enabling the United States to grow into one of the world's superpowers in a short amount of time. As with the new private companies in space, the transcontinental railroad brought down the cost of travel. It cost $1,000 to travel across the United States prior to the railroad being built, a figure that dropped to $150 after construction was completed.[6] Where a station was built, a town would flourish, and train tracks also enabled the movement of goods across large distances, establishing a major advantage for the industrialized North during the American Civil War. But to say this was all due to the private sector's money-making endeavors doesn't tell the whole story.

"Everybody likes to point to the railroad and say that, 'Oh, well back in the nineteenth century, when all this was all being built up, it was all built by the private sector.' Well, hold on a minute," James Vedda, senior policy analyst for the Center for Space Policy and Strategy at the Aerospace Corporation, told me.[7] "They didn't do it alone because they were given huge amounts of land to lay their tracks and to build their stations. And not just a little strip of land wide enough for the tracks, they were usually given up to a mile on either side."

The land grants were given to these companies by the U.S. government as part of the Pacific Railroad Acts of the 1860s.[8] This made them a great early example of a public-private partnership. "I read one estimate that in the nineteenth-century development of the railroads, the railroad companies were given land grants that if you total them all up together were equivalent to the size of Texas," Vedda added. "They sold off all that extra land [and] they found that they got to keep the money. Besides that, the U.S. Geological Survey went out and did this surveying for them and gave them the results for free so that is a significant cost that they didn't have. So that was a great example of a partnership."

In space, SpaceX and Blue Origin are the railroad companies, building a pathway into the beyond that others can use. The next wave of entrepreneurs are those who set up hotels, bars, and factories in places with easier access and increased populations. The private launch companies have worked closely with NASA and the government to get as far as they have. Obviously they haven't been given any land in space, and that's perhaps the greatest flaw of the analogy. But there are lessons to be learned regardless. The transcontinental railroad brought great economic prosperity to America, but not without a cost. The government gave away a huge amount of land that belonged to Native Americans, displacing them and also damaging wildlife. While there may not be any natives in space (that we know of), this colonial attitude can be dangerous, as I will explore in many chapters of this book.

It could be that only capitalism has the potential to put a permanent base on the Moon or people on the surface of Mars, but if that's true, we need to ensure rules, regulations, and best practices are in place well before we reach these milestones. The startup world has long raged against such regulations, but these libertarian attitudes simply won't work in the harsh environment of space, where safety and freedom are far from guaranteed. While nobody wants international regulation to limit innovation, there must be a balance.

To understand the commercial space industry, we must first look at its short history and the landmark moments that moved it from a frowned-upon concept to the bastion of progress many see it as today. The first chapter will take a look at the early promise of commercialization, from the moment Russian cosmonauts and American astronauts shook hands in space, and the man in the 1980s who struck the first major blow against government monopolies. In the second chapter we will follow efforts to privatize space that began as early as the 1990s. The early 2000s ushered in a new business model in space: selling tickets on rockets. Space tourism and other early investments during this period offered a proof of concept for would-be space entrepreneurs. The end of the first section will see the building blocks of space tourism put in place, with a major shift in the relationship between NASA and private companies as the backdrop.

Then we turn our attention to today, where money pouring into private space companies is being used to buy political influence, which raises a series of ethical issues. The resurgence of off-world tourism has the potential to democratize space, but the prices are still way too high for anyone but the one percent, which undercuts how much it can affect the world in a positive manner. The cluttered nature of lower Earth orbit also poses immediate danger, and sadly reflects the lack of care humans have for their surroundings here on Earth. Technology is shrinking and becoming cheaper, making it easier for companies to add their own hardware to the orbital traffic jam. But one country in particular threatens the status quo more than others—China.

Last, we'll look further into the future, where it's hard to differentiate between concepts that sound like science fiction and legitimate plans. For that section, we will start close to the Earth with private space stations, where four proposals are on the table, all of which are planned for the next decade or beyond. The different forms of fundraising each of these companies employs reveals much about the state of the industry. Going beyond the orbit of the Earth is the ultimate goal for many space enthusiasts, but if humans are to survive long journeys in space, they may need to make certain adjustments. This may involve drastic measures—and new industries—like gene editing to make a human perfect for space travel, with far-reaching consequences. Once problems like these are solved, space companies can properly turn their attention to outer space, starting with the Moon. It now seems very likely the next human on the Moon will either be part of the Chinese space program, or sent there via a commercial spacecraft. The reason for the commercial rush to our lunar cousin is one humanity is well accustomed to—a dash for resources. Finally we'll reach Mars, a planet with an obsessive fan base already established here on Earth, and the subject of the grandest plans by commercial space companies, even if some of them have already failed.

When humans do reach Mars—and I firmly believe we'll reach the Red Planet in my lifetime—they will face some tricky decisions regarding an agreement signed way back in 1967, which is still the primary governing treaty in outer space today. In its time, the document was visionary, but it was also intentionally ambiguous, and the debate over what it allows and what it forbids is heating up.

PART I

THE PAST

1

The Cold War and
the Outer Space Treaty

THE QUESTION OF WHAT ONE CAN AND CAN'T DO IN SPACE HAS evolved since the first space race began nearly 70 years ago, and the answer has depended on the era in which it is asked. If international space law at the dawn of the space race sought to restrict militaristic buildup and nuclear dominance, the era we live in now seeks to encourage private corporations to enter a realm once strictly controlled by government-funded programs. That shift could begin only after the United States and the U.S.S.R. ended their century-defining extraterrestrial rivalry.

On July 13, 1975, two spacecraft were launched from opposite ends of the world with the intent to rendezvous in orbit for a joint interplanetary mission.[1] From Cape Canaveral came an unnumbered and unused vessel from the Apollo program that had ended three years prior. From Russia, it was Soyuz 19, then the latest in Russia's Soyuz fleet, which operates to this day. Online, you can watch rough footage of each launch and, because each craft was outfitted with a small camera, the docking of both vessels in orbit, with France in the background looking up from the blue planet below.

Greeting each other through nozzle-like corridors, the crews embraced in bear hugs and exchanged diplomatic offerings. A prepared statement of congratulations by General Secretary Leonid Brezhnev was read aloud, which stated: "The Soviet and American spacemen have gone into outer space for the first major joint scientific experiment in the history of mankind. They know that from outer space our planet looks even more beautiful. It is big enough for us to live peacefully on it, but it is too small to be threatened by nuclear war."[2] President Gerald Ford rang the crew directly from the White House, and said somewhat prophetically, "The day is not far off when space missions made possible by this first joint effort will be more or less commonplace."[3]

A great deal of preparation between the two world powers went into creating this moment, and the show of goodwill was on full display. Leading up to the launch, engineering teams had been allowed to inspect each other's spacecraft, making for an unprecedented level of access to what are often technological secrets. A large embroidered patch stitched onto the uniforms of all crewmembers bore the words "Apollo" in blue and "Союз" (Soyuz) in red surrounding a Pan Am–like globe. The teams cohabitated in space for two days and performed experiments such as maneuvering the spacecraft to create an artificial eclipse of the sun to photograph its corona. But perhaps the bigger accomplishment occurred upon reentry, when the safe return of both spacecraft to Earth, mission accomplished, marked the end of the space race, or at least of the era of two world powers competing for dominance in the cosmos.

At the time, the United States and the U.S.S.R. were still technically engaged in a Cold War, a bloodless conflict fought with surveillance, espionage, and technological competition. It didn't end with the ceremonious denouement of the space race, but that certainly helped. Nationalistic interest in space, however, all but dried up after Apollo-Soyuz. Gone was the patriotic fervor that had brought us from sending satellites into orbit to landing people on the Moon in only a decade. Gone too was any collective achievement that all of humanity could champion. The urgency had disappeared from space travel.

The two nations went on to develop their own independent and much quieter space station programs. While the Soviets began a program of launching stations into orbit at a rate of one every two years, the United States launched one single space station, Skylab, that went largely unmanned for the duration of its orbit due to safety concerns.[4] Rather than try and salvage Skylab, NASA cut its losses and turned its attention instead to developing a space shuttle program that could bring crews to space and land them safely back on Earth in the same vessel.

Back then, the United States and U.S.S.R. were the only two nations capable of sending anything into orbit, and breaking that monopoly wouldn't be easy. Yet challengers were emerging: Only two months prior to the Apollo-Soyuz mission, in May 1975, two space programs—the European Launcher Development Organisation and the European Space Research Organisation, which had both been founded by a group of Western European nations—kicked into high gear when they merged to become the European Space Agency (ESA), which to this day remains the EU's primary space program.

These newcomers had to adhere to the same rules written to keep the two sparring Cold War nations in check. When the United States and Russia began to make progress in their space exploration efforts, many feared what would happen if the space race were to turn hostile, or if either country were to weaponize space. At the height of international mistrust, a U.N. committee managed to draw up an agreement that still dictates what can and cannot be done in space.

The Outer Space Treaty (OST) is the most important legal framework of international space law. It was set forth by the United Nations in 1967, but its origins trace back to the advent of the space race in the mid-1950s. That was the heyday of space heroes Buck Rogers and Flash Gordon, and novelists Ray Bradbury and Isaac Asimov were at their height, buoyed by America's real hopes for human space travel.

In March 1952, *Collier's*, a national weekly magazine published an issue devoted to the plausible cosmic future of humanity. "*Mankind Will Conquer Space Soon*" read the cover, with an illustration of something resembling a stealth bomber firing off a glowing hot honeycomb over a planet that is half Martian, half Earth.[5] While these articles took great imaginative liberties—this was five years before the Soviet Union sent the much more primitive looking Sputnik 1 satellite into orbit—the *Collier's* series was one of the first attempts by scientists and journalists to influence the opinions of lawmakers and the general public about space. And the questions they asked were concrete and forward-looking, covering topics such as the ownership of space, how maritime laws can be applied off-planet, and even speculation about establishing space stations that can refuel rockets and spacecraft on the way to more distant parts of the solar system.

At the time, the most immediate concern of space wasn't *when* will we get there as a species, but *who* will get there first as a nation. "What you will read here is not science fiction. It is serious fact," one article begins. "Moreover it is an urgent warning that the U.S. must immediately embark on a long range development program to secure for the West 'space superiority.' If we do not, somebody else will. That somebody else very probably would be the Soviet Union."

As early as 1954, both nations had begun developing artificial space satellite programs, but it wasn't until 1955—when President Dwight D. Eisenhower announced via his press secretary an endeavor called Project Vanguard, in which the United States planned to launch the first satellite, Explorer 1, into orbit by 1958—that a real target was set.[6] Publicity for Project Vanguard grew, and it might possibly have succeeded in sending the first satellite into space had it not been for two mishaps.

The first was that the Russians suspected foul play and made it known that such developments could be interpreted as a buildup of arms in preparation for a militaristic strike. Hoping to avert a crisis, at a Geneva Summit soon after Project Vanguard was announced, President Eisenhower discussed with Soviet premier Nikolai Bulganin (essentially the Soviet vice president) a proposal he called Open Skies, a peace treaty in which the two nations would be permitted conduct surveillance flights over each other's territory to ensure that neither country was preparing for attack.

But trust between the two nations, while never robust, was rapidly disintegrating, and the proposal stalled out under accusations that Open Skies was nothing more than a masked attempt to learn more about Soviet aerospace and ballistic technology.[7] While there was some truth to that claim, neither party was willing to push the discussions and risk further escalating tensions.

The second mishap was time, or rather how much time the United States took to get Project Vanguard off the ground. The Vanguard rocket, while it looked sleeker and more appealing as a space rocket to the media, was not quite as powerful as the rocket initially proposed for the launch, the more missile-like Redstone. While the United States tinkered with Explorer 1—the satellite the Vanguard rocket was to bring into orbit—the Soviets were busy with their own satellite project: Sputnik.

Though the Sputnik project was known to the United States, its launch date was not, and in October 1957, the Soviets managed to launch Sputnik 1 into orbit, three months before the United States got around to launching Explorer 1.[8] To add insult to injury, less than a month later, while the final preparations for Explorer 1 were still being made, the Soviets followed up with a second satellite, Sputnik 2, which sent the first living earthling, a dog named Laika, into orbit. The American media wasn't kind. "A wisecrack much repeated in the U.S. last week was that Project Vanguard, the U.S.'s earth-satellite program, ought to be renamed Project Rearguard," declared *Time*.[9] The magazine rightly perceived that these barbs "clothed in humor the widespread feeling of resentment stirred up by the Russians' great cold-war propaganda victory."

In January 1958, the United States caught up with Explorer 1, but by then, a widespread fear known as the Sputnik crisis had risen among Americans who fretted that the Russians were in control of vastly superior technology. The same year Explorer 1 was launched, a federal agency devoted to space exploration was formed: the National Aeronautics and Space Administration (NASA).[10] As firing rockets off into space could be particularly difficult to conceal, both sides used space exploration efforts to conceal more cloak-and-dagger surveillance projects. That's when the United Nations stepped in.

In 1958, a committee was formed to explore how best to keep space a neutral and peaceful territory. Organizing under the official name of the U.N. Committee on the Peaceful Uses of Outer Space (COPUOS), its stated mission was to "review the scope of international cooperation in peaceful uses of outer space, to devise programmes in this field to be undertaken under United Nations auspices, to encourage continued research and the dissemination of information on outer space matters, and to study legal problems arising from the exploration of outer space."[11]

As the Cold War, particularly its extraterrestrial front, escalated in the '60s, COPUOS felt mounting pressure to produce a hard document that would be internationally ratified. But it wasn't until 1967 that the committee was able to establish a document tongue-twistingly titled the Treaty on Principles Governing the Activities of States in the Exploration and Use of Outer Space, including the Moon and Other Celestial Bodies, before being shortened, at least in reference, to the Outer Space Treaty (OST).

Among its principles, the OST bars its signees from placing weapons of mass destruction in Earth orbit, installing them on the Moon or any other celestial body, or otherwise stationing them in outer space. It exclusively reserves the use of the Moon and other celestial bodies for peaceful purposes and expressly prohibits their use for testing weapons of any kind, conducting military maneuvers, or establishing military bases, installations, and fortifications.[12] However, the treaty does not prohibit the placement of conventional weapons in orbit and thus some highly destructive attack strategies, such as kinetic bombardment—in which an object is dropped from orbit, gaining very high speed and causing enormous damage upon impact—are still potentially allowable.

The treaty also states that the exploration of outer space shall be done to benefit all countries and that space shall be free for exploration and use by all states. The treaty explicitly forbids any government from claiming a celestial resource, such as the Moon or a planet. Article II of the treaty states that "outer space, including the Moon and other celestial bodies, is not subject to national appropriation by claim of sovereignty, by means of use or occupation, or by any other means." However, the state that launches a space object retains jurisdiction and control over that object. The state is also liable for damages caused by its space object.

In January 1967, the United Kingdom, Soviet Union, and the United States signed the document, along with 57 other countries ratifying it, and it went into effect in October of that year. Countries can withdraw from the treaty a year after a written intention is delivered to the three countries that first signed. Amendments must be approved by a majority of states, and then are only binding for the countries that accepted them. While it didn't stop the spy satellites or slow down the space race, it did ensure the world wouldn't be ended by a nuclear bomb being dropped from the Earth's orbit. U.S. president Lyndon B. Johnson described the signing of the treaty as "an inspiring moment in the history of the human race."[13] It remains the only document that protects human interests in space and it was soon put to the test.

In 1969, just two years after the treaty was signed, Neil Armstrong and Buzz Aldrin landed on the Moon and planted an American flag on the surface. This immediately brought one article of the treaty into sharp focus. Article II of the Outer Space Treaty reads: "Outer space, including the moon and other celestial bodies, is not subject to national appropriation by claim of sovereignty, by means of use or occupation, or by any other means."[14] However, before the flag even left the planet, the Americans made their intentions clear, announcing that they were in no way challenging this part of the treaty. After the incident, the American government released statements and even passed a law appeasing the international community that the flag merely represented the achievement of reaching the Moon, rather than claiming any kind of ownership of the land. In the '60s, the treaty aimed to stop a messy, dangerous land grab in space, and it worked.

But there were fewer players back then, and it was national and not private endeavors under the treaty's purview. This all changed in the mid-'80s, and once again the notion of who controlled space was changed with the launching of a satellite.

* * *

ALMOST ALL OFF-WORLD ACTIVITY WOULD REMAIN UNDER the aegis of the United States and the U.S.S.R. well into the 1980s, with only the ESA making any other significant inroads into space exploration. ESA launched a high-orbit telescope with NASA in 1978, and launched its first deep space exploration mission, Giotto, in 1986.[15] It wasn't just the huge expense of sending aircraft and people in space, but the space industry was also under tight government control. Nowhere was this more readily enforced than satellites. A global consortium of governments called the International Telecommunications Satellite Organization (Intelsat) had been responsible for all international TV and telephone satellite transmissions from 1964 onward.[16] That is, until one headstrong individual with a large personal fortune broke the government stronghold and carved the path for private enterprise in space that so many companies have since followed.

Reynold "Rene" Anselmo was not the kind of person you'd expect to take on the space industry, but perhaps he was best suited for the job. Born in Bedford, Massachusetts, in 1926, Anselmo was a scrappy and pugnacious teenager who, in 1942, at the age of 16, lied about his age to join the military, becoming a tail gunner of a dive bomber, no doubt spurring his interest in aeronautics.[17]

After the war, Anselmo moved to Mexico, where he got involved in the entertainment industry, producing and directing theater and television shows. He worked on translating American productions into Spanish and befriended Emilio Azcárraga Milmo, the son of the head of Mexico's largest media and television company Televisa. By 1954, Anselmo was marketing Televisa's shows to other Latin American countries, and in 1963, he moved to New York to become president of Spanish International Network, known as SIN. The Spanish-language network thrived in the United States, becoming a major presence, and Anselmo was heavily invested in the network's holding company and several of its TV stations. But in the 1980s, SIN came under scrutiny after competitors alleged it was controlled by Anselmo's former company Televisa. The Federal Communications Commission (FCC) was brought in to investigate, as U.S. laws dictated foreign entities were allowed to own a maximum of only 20 percent of a domestic network. Eventually the company was sold to Hallmark Cards in 1986, and Anselmo reportedly walked away with $100 million.[18]

Even before selling SIN, Anselmo had set up his next venture, and it was more ambitious than anything he'd previously undertaken. Anselmo founded PanAmSat—a satellite service catering to news, entertainment, and telecoms companies—in 1984. That same year, he applied to the FCC for permission to launch a private satellite, an activity that was completely con-

trolled by Intelsat, jointly owned by over 100 governments, of
which the United States had a 25 percent share. The Intelsat
consortium built and operated the satellites, while the marketing was done by individual member-owners. Under the
international agreements in place, the participating countries
designated companies, which were usually government-owned
telecommunications firms, to serve as their representative to
Intelsat. Those companies would transmit and receive data from
Intelsat satellites and charge their customers, who were supply-
ing television, phone, and data services to everyday consumers.
The organization was headquartered in Washington D.C., and
the laws were set up in favor of Intelsat.

"It was a government consortium which controlled international satellite communications. It was inefficient. The cost
was very high," NanoRacks CEO Jeffrey Manber, who worked
under Anselmo at PanAmSat at the time, recalled to me.[19] "I was
a young man, and I had the privilege of working with him, the
president of PanAmSat, and he wanted to send soap operas from
Mexico to America, or America to Mexico, I forget which one.
And he found out it was incredibly expensive. So he said, okay,
I'll put up my own satellite. Well, he wasn't allowed to own it."

Anselmo's response was to go up against an international
consortium controlled and based in the United States and win
on its territory. According to some reports, he countered the
superior lobbying power of Intelsat by, among other strategies,
sending lawmakers in the U.S. letters featuring his dog Spot.[20]
Eventually, the FCC relented and granted Anselmo permission
to launch a satellite and operate it outside of Intelsat's control.

Anselmo was wealthy, but no billionaire, so when he put
together a plan, the budget was small. He managed to keep costs
down by taking calculated risks and shopping wisely. He had
bought a satellite from RCA Astro Electronics, paying a cut rate
because it was built for another customer, which had canceled
their contract. It's estimated that around that time a full-price
satellite would have cost around $80 million, but Anselmo paid

just $45 million.[21] Although the Reagan administration first showed support for limited competition with Intelsat in 1983, it took until September 1987 to get PanAmSat's final launch approval from the FCC. Anselmo now had two problems: he needed someone to launch the satellite into space at an affordable price, and his business case was still weak, considering only Peru had agreed to allow its population to communicate with his satellite.

Over the next year, Anselmo patiently negotiated with other governments one by one to secure "landing rights," which allowed him to sell the satellite's services to companies operating in those countries. After Peru, the first to agree to a deal was West Germany, followed by Margaret Thatcher's United Kingdom, according to Manber, who described the fact the United States wasn't first as an "embarrassment." By the time of the launch, Anselmo had agreements with six countries.

The Challenger disaster had occurred in January 1986, so NASA's space shuttle program was suspended. Anselmo managed to secure a cheap ride into space with the European consortium Arianespace, which was suffering something of a downturn in its reputation as four of its rockets had blown up since 1981. Arianespace was debuting a new series of rockets, and Anselmo was gambling on their success. He booked a spot aboard the maiden voyage in June 1988, essentially a test flight. According to the *New York Times*, Anselmo paid just $9 million for the flight and insured the satellite for only half of its replacement value, to cut even more costs.[22]

His gamble paid off, and the satellite was launched into orbit without a hitch. By the early 1990s, the satellite was in almost constant use and offered much-needed competition to Intelsat's dominance. Other companies began to make plans to launch their own satellites, and the monopoly was broken. Anselmo later admitted to the *New York Times* that his gamble was based on a hunch rather than any market data he'd procured. "My theory," he said, "was that I couldn't imagine putting a satellite up there and offering all this technology without it being used."

The satellite helped U.S. news stations aggregate news from all over the world. One of PanAmSat's first customers was CNN, which used the satellite to send programming abroad and to receive material from its bureaus overseas. It was also used by European news organizations like the BBC and Sat.1, a German network, to broadcast live from Washington D.C., while other American networks used it to send material abroad and supplement links to the Middle East.

Anselmo's crusade against an international consortium of governments unlocked part of the commercial space industry, and showed the sort of spirit needed to break into markets controlled by governments. The maverick CEO proved the stranglehold of governments in space could be broken, which led to the thousands of private satellites in the lower Earth orbit today. The precedent had been set—companies can make money in space, and they don't need the government's help.

As Anselmo was challenging Intelsat over their satellite monopoly, prominent figures in both the United States and Russia were advocating for the private sector to take a larger hand in matters in space. Somewhat surprisingly, it was the Russians who made the first real move, and just like with Intelsat, Manber was right in the middle of it again.

2

The 1990s: A False Dawn for Private Interests in Space

IN 1991, THE SOVIET UNION COLLAPSED, AND WITH IT THE Soviet space program. As Russia rebuilt, the government tried to ensure the country still had power in space. To do so, the Russians looked to do something that even the home of the free market had not yet managed—to privatize space. It was a bold move from a brand new country, and one that embarrassed American space entrepreneurs and had them ready to jump aboard the Russian efforts.

After the fiercely competitive space race of the 1960s, culminating in the Moon landing, the Russians and Americans moved the goalposts in the '70s, and '80s, both attempting to establish space stations in lower Earth orbit. In the '80s—after a decade of experiments with space stations sent up by both countries—the Soviet Union launched Mir, the first modular space station.[1] It was larger and more advanced than any previous station, surpassing anything the Americans had accomplished. It was also a chaotic, supremely successful project that served as a testament to the sometimes risk-taking nature of the Soviet space program. Most importantly, in hindsight, it would host some of the earliest forays into the privatization of space.

The base block of Mir was launched on February 20, 1986, and received its first residents in May that year, when Leonid Kizim and Vladimir Solovyev stayed onboard for about two months. In the following years, the Soviets added to the base module and the station became shaped either like a prickly hedgehog or a dragonfly with outstretched wings, depending on who was describing it. Mir was a trailblazing example, and the Americans wanted their own version. So they developed a program called Space Station Freedom, at the time the largest international scientific and technical cooperative program in history, according to the International Organization for Standardization.[2] The United States was joined by Japan, Canada, and several member countries of the European Space Agency in developing the project, which was announced to the world in Ronald Reagan's 1984 State of the Union address. The goals of the project feel familiar today—to provide an observation post for astronomers, a laboratory for scientists, and a factory for companies.

Mir may have been the most advanced space station ever, but toward the end of the 1980s and early 1990s, the program faltered as the Soviet Union began to collapse. After the 1991 dissolution of the union, the newly formed Russian space agency Roscosmos took over management of Mir.

The end of the Soviet space program raised new issues for the Americans. There were immediate concerns that space experts from the Soviet Union would begin working for rogue states and the technology to launch satellites and space stations into orbit could be in hands of America's new enemies, such as Iran and North Korea. To combat this, the United States began to reach out to the Russians to discuss possible cooperative space missions. These talks were attended by Jeffrey Manber, who had since left PanAmSat and had begun working for the Russian space company Energia. Manber had been drawn there by the efforts made in the dying days of the Soviet Union to commercialize the space program. "It was a time when two macro events were coming together," recalls Manber.[3] "The first one was the very pro-business attitude of the Reagan administration and the second was the collapse of the Soviet Union. Under Prime Minister Gorbachev, the Soviets tried to monetize the few assets that they had that were world class and in demand. So they privatized Aeroflot, the airline; they privatized the Bolshoi Ballet; and they sought to privatize space—the space program. For us in the industry, it was an extraordinary radical and revolutionary move that brought space for the first time into the mainstream of business."

As part of the privatization movement, Manber had helped secure a deal for an American pharmaceutical company to conduct drug research on the Mir space station, and the Russians had asked him to help introduce them to other international companies that could become clients.[4] At this time, George Bush was president, and Manber secured the blessing of the administration beforehand. "It wasn't permission, but I got in writing a letter that said: 'Jeff, good luck on this exciting adventure to work with the organization Energia,'" said Manber. "And so at least it was on paper. If things went a different way, I wouldn't be hauled before Congress or something like that."

So Manber took his seat on the opposite side of the negotiating table from his country of birth. "I started working with Energia and they understood about space stations. They understood about cost efficiencies, and they were on a commercial pathway. And so as a result of that, I had the pleasure, the honor, I suppose, of carrying over the first contract between NASA and the Russians. And that was to use Soyuz as an escape vehicle for Space Station Freedom. I was there as the Russians suggested to the Americans that Freedom should never be built, could never be built, which was a view echoed by NASA administrator Dan Goldin. And they suggested what became the International Space Station. And so I was there at the beginning of what is today, the International Space Station. But I was on the Russian side."

While they agreed to cooperate and build the International Space Station (ISS), the two countries also struck a deal for

NASA's space shuttle to travel to the Mir space station, carrying both Russian and American astronauts. During these missions, stories began to emerge from Mir, including how messy, unkempt, and downright dangerous it could be. Astronaut Jerry Linenger, who became the first American astronaut to conduct a space walk from a foreign-built space station in 1997, was quoted on the NASA website saying Mir was similar to "six school buses all hooked together.[5] It was as if four of the buses were driven into a four-way intersection at the same time. They collided and became attached. All at right angles to each other, these four buses made up the four Mir science modules. . . . Priroda and Spektr were relatively new additions . . . and looked it—each sporting shiny gold foil, bleached-white solar blankets, and unmarred thruster pods. Kvant-2 and Kristall . . . showed their age. Solar blankets were yellowed . . . and looked as drab as a Moscow winter and were pockmarked with raggedy holes, the result of losing battles with micrometeorite and debris strikes over the years." Linenger had to endure the most severe fire ever reported on a space station orbiting the Earth and several system failures.

Despite the eccentricities of the Mir space station, the partnership between the American and Russian space programs was declared a success. The station's commercial interests were controlled by Energia, where Manber worked "on marketing the Mir," he said. "We worked with pharmaceutical companies. We worked with Israeli milk companies. We worked with Pepsi-Cola, we worked with RadioShack. There was total disdain and humor and mocking from many in the American space industry. And at the same time, I was extremely proud of the Russians."

Two countries, one emerging from a collection of communist states, and the other a bastion of capitalism, had considered commercializing their space efforts, and it was Russia who made the most headway, to the derision of their counterparts in America. The Russians were forced in this direction by economic factors, but Manber still finds it embarrassing that the United

States didn't make the first move. By the mid-'90s, he wasn't the only member of the space industry frustrated by the efforts of NASA. The agency was struggling to live up to the achievements of the late 1960s and '70s, and innovation was held back by politics. Ideas that we celebrate today as originating from the private sector—like reusability in rockets—have been around since the 1980s, but NASA had dismissed them. Bill Stone was one of those people with grand ideas who felt betrayed by the agency. Engineers like him were working hard to make space travel as efficient as possible, but efficiency jeopardized a major driver of U.S. efforts in space—job creation.

* * *

STONE WORKED WITH NASA FOR DECADES, AND HAS EXPERI-enced firsthand the frustrations that have prompted calls for the private sector to step up and into space. The ideas he pitched to NASA about increased reusability in rockets fell through the cracks, deemed at once too expensive, ambitious, and worst of all—potential job killers. In 1989, Stone was in the second of three groups vying to become the next batch of NASA astronauts.[6] He wanted to return to the Moon for his country and become the first human to step foot on the lunar surface since 1972. But during training he was called into the office of the head of the Johnson Space Center and given some disappointing news. He was told that should he be selected, he would not be going to the Moon, and neither would any of his fellow astronauts. Missions to the Moon were no longer among NASA's goals, as the agency focused on unmanned missions further out in the solar system and building and maintaining the ISS closer to home. "It was my first dose of realism in the space world," Stone told me on a Skype call, during which he talked nonstop for more than an hour, pausing only when the connection failed briefly.

But that wasn't the end of his disappointment. He'd also been working on an idea that had the potential to revolutionize space travel. From 1980 onward he was researching a way to reuse the external tank of a space rocket. Among NASA's most limiting and longstanding problems is its inability to reuse many of the components on a rocket. Nowhere is this made clearer than when watching NASA promo videos in which gigantic orange containers are jettisoned from rockets and shuttles right as they achieve orbital velocity. Those capsules wastefully crash back down to Earth as garbage. To Stone, the plight of the external tank represented the waste of the space industry. This was a 65,000-pound object being used to store the necessary fuel to get the shuttle off the ground, but immediately discarded once it had done its job. Each of these external tanks were about 20 times bigger than the ISS, which would be launched in the '90s, according to Stone, and his initial idea was to safely deliver the tanks back to Earth, where they could be repurposed as industrial lab space. But Stone realized there was an even more ambitious and potentially lucrative use of the external tank—as a fuel depot.

Storing fuel somewhere else besides here on Earth is a notion under strong consideration today, but back in the 1980s this was a fairly new concept. The idea is simple—every extra pound on board a flight raises the cost of going to space considerably, and one of the heaviest things to carry is the fuel required for the mission. If the fuel could somehow be put in orbit, shuttles, probes, and other spacecraft could blast off lighter, collect fuel while orbiting Earth, and go on their way to distant worlds with the fuel they need. But just as Stone began to make progress, he encountered familiar bureaucratic issues. This time, it would change his thinking forever.

Stone was working with a group of around ten people, looking at the possibility of putting an external tank in orbit, and had reached the point where a flight mission had been planned. As with all NASA work, one of the final stages of green-lighting a mission involves taking the proposal to the White House, where space endeavors are put under the scrutiny of the least scientific of processes—politics. The year was 1992, and Stone was called into the director's office at NASA.[7] When he sat down, the director told him he had "bad news and good news." Stone, a forthright, matter-of-fact person, speaks with little emotion, but it's clear this is a recollection that bothers him more than others. "I said okay, give it to me," he recalled. Stone was told that his flight proposal was being shot down on the basis of politics. "Politics?" Stone asked. "This is engineering." No, the director replied, the project was being viewed as a threat to the jobs program the space industry enabled. The test flight was being rejected because it had the potential to make it cheaper and more efficient to put objects into space, and this would have resulted in a loss of jobs in the aerospace industries.

As Stone was told that his project was being shelved for political reasons, he was also given a research grant to look into anything he wanted within reason. He chose automation, as he thought robotics could play a key role in future space exploration. "We were running automated vehicles at 100 kilometers per hour across the campus with nobody at the wheel, a decade or more before Google and others got into this," he said. But he hadn't forgotten about the external tank, the huge under-utilized cavernous space attached to the back of the shuttle. He considered a new question: What would it take to put the external tank on the surface of the Moon? "We ran the calculations and the stunning answer that came back was not very much," Stone enthused. "You could actually probably privately raise that money if you had the tank in orbit, go up and retrofit it, put guidance propulsion systems on it, and you could break off the front—the pointy section of it—and use the inter-tank structure as the mounting points for your landing systems and your propulsion systems and your guidance systems."

Stone had gathered the expertise from his automation work and believed the plan could succeed. But he realized that to attract investment from the private sector, he couldn't just send a tank to the Moon, he had to bring something back. He started working on the calculations. It still cost a huge amount of money to get anything to the Moon, even without any people onboard, and he needed something valuable enough to sell back on Earth and achieve a positive return on investment. The numbers didn't add up. "We couldn't bring gold bricks back from the Moon and make money because of the transport costs, and largely this had to do with the way in which things get launched into space."

Stone abandoned plans to bring anything back to Earth. The only materials which could potentially be mined and sold at a profit, after factoring in the lofty transport costs, were rhenium, one of the rarest elements in the Earth's crust, and rhodium, an extremely valuable member of the platinum group. But there wasn't any solid proof that either element existed on the Moon to be mined. His next idea involved gathering resources and placing them somewhere besides the Earth. One mineral that had been proven to exist on the Moon was ilmenite, which was discovered during the Apollo missions. Ilmenite isn't exactly valuable, but it is made up of around 60 percent oxygen. Bringing oxygen back to Earth would clearly be a thankless task, but it does play a significant part in space travel. Oxygen isn't a fuel itself, but it's required to help with the combustion of whichever fuel is used. That means large quantities must be onboard when the rocket takes off. This is all extra weight to get out of the Earth's atmosphere, so Stone proposed placing an oxygen depot in the Earth's orbit at which spacecraft could stop off and replenish their oxygen levels prior to going deeper into space. Again, he ran the numbers, and although he was significantly closer to making a profit, it still fell short. Other companies researched this area as well, but none were able to make the numbers work. Discouraged by his inability to turn a theoretical profit, Stone returned to his work on automated vehicles, until another example of governmental failure and a completely chance meeting led him to a light bulb moment years later.

In September 1995, *National Geographic* published an article about an expedition into the Sistema Huautla in Oaxaca, Mexico, the deepest caving system in the Western Hemisphere. The article had been written by Stone, who led the expedition. Stone recounted the thrill of pushing into new sumps and chambers and grimly described diving into a flooded vein to retrieve a drowned colleague. While this was all thrilling, what captured the most attention was the suit Stone used to dive into the water and locate the body—a wetsuit equipped with a breathing system that recycled its own air tanks, allowing wearers to breathe underwater for over 24 hours.

After the article was published, Stone received a host of invitations to give talks about his adventures and expertise, which he turned down—except for one. A Boy Scout leader had managed to secure permission to go caving in a system in Western Virginia that Stone had visited in his youth but had since been closed off to the public. Sensing an opportunity, Stone arranged for his two sons to join him and accepted.

About a week later, Stone and his two sons were in a cabin in West Virginia cooking breakfast for and gearing up with nearly 50 Boy Scouts. During the expedition that followed, the scout leader, an Air Force colonel named Dale Tietz, struck up a conversation with Stone asking what he did when he wasn't exploring caves. Stone mentioned his background in aerospace engineering and some of his undeveloped ideas, including an unfunded project to develop reusable rocket fuel tanks.

Much to Stone's surprise, Tietz followed up with questions of greater and greater detail, phrased in space jargon, commiserating with Stone about NASA's wasteful expenditures. At some point on the trip, Stone finally asked exactly what Tietz did for a living. As it turned out, the man organizing caving trips for the Boy Scouts in Virginia was a high-ranking official at the Pentagon who was working on the remnants of the space segment of the U.S. government's Star Wars program.

"We're talking a $2 billion-a-year program that nobody knew about that was taking place in orbit. And he was in charge of it," said Stone. So at the end of that week, Stone drove to Alexandria, Virginia, to what he described as "one of the seedier parts of town" and "not a high-rent area." Tietz met Stone on a dilapidated block and walked him to one of the older, more weathered buildings. Tietz pulled out a key and ushered him inside. They walked down a wood-floored corridor for about 10 meters, when Stone saw something that didn't quite belong—a stainless steel door with a cyber lock on it. Tietz punched a series of numbers into the lock and opened the door to reveal mission control for Clementine, a joint space project between the Ballistic Missile Defense Organization and NASA.[8]

"It's just like Johnson Space Center mission control, computer screens on all the walls. There's 20 or so guys in there running various parts of the mission," explained Stone. The Clementine mission had two major objectives: to test how sensors and components of spacecrafts fared when exposed to extended periods in space, and to make observations of the Moon and a nearby asteroid. When Stone walked through that door in 1995, the mission had already run its course and all that was left was to sift through the data. But by chance the team behind the project had been given the opportunity to collect more than they'd originally planned.

A thruster malfunction while the spacecraft was attempting to leave lunar orbit meant the second part of the mission, to the asteroid, was not possible. Clementine was put into a makeshift orbit around the Moon, and the team behind the project began to think what extra science they could carry out in these circumstances.

According to Stone, it was a man named Dr. Paul Spudis, the deputy leader of the mission's scientific team, who decided to use the radar on the spacecraft to bounce a signal off the surface of the Moon and pick up the reflected wave at the Goldstone Deep Space Communicatons Complex in California. What they found was quite odd, and completely revelatory to Stone.

"Colonel Tietz takes me in, and he goes over to Paul Spudis and says, 'Paul, please bring up Track 134.' And so here's this computer screen. I'm looking at a graph, an x, y graph. And so there's the signal going across the bottom and about halfway across the screen, suddenly this thing spikes way the hell up and then comes back down, and then keeps sputtering around the low end," he recalled.

Stone was naturally curious at what he was looking at, and Dr. Spudis duly explained. One part of the graph was the latitude measuring, and the place where the spike occurred was at minus 90 degrees—the south pole of the Moon. The spike itself indicated the presence of hydrogen. By this point, Colonel Tietz was watching Stone and smiling. "I said, 'How much hydrogen?' And he said, 'By my calculations, no less than ten to the ninth billion metric tons of hydrogen.' Okay, so 10 billion tons of hydrogen," Stone said. This isn't just hydrogen, of course, it's probably in the form of water or ice and there may be other things present in it, but it still represented a large helping of a material Stone could find very useful. "I said, 'You just completed the puzzle,' and at that point it all fell into place."

Stone hatched a plan to collect hydrogen from the Moon using automated mining machinery, send it up into the Moon's orbit, and create a refueling station for ships heading out into the solar system. Despite the breakthrough, Stone's plans would take an enormous investment to realize, meaning he was automatically priced out of NASA. Because the commercial sector of space had yet to take off in the United States, he had nowhere to turn.

It wasn't until 2007 that Stone started Shackleton Energy with the purpose of implementing his hydrogen mining idea, and even now he is on the lookout for a cool total investment of $17 billion, meaning he's holding out for a mega-billionaire, an oil-rich country with a single ruler, or a gigantic corporation for investment. Back in 1995, Stone had no chance of developing the idea that had the potential to increase the affordability of space travel considerably and accelerate private space endeavors. As the decade drew to a close, the prospect of a thriving American commercial space industry was still distant. But in Russia, Manber was about to make a large bet on taking an old piece of infrastructure private.

* * *

AROUND 1997, MANBER GREW TIRED OF THE POLITICS BETWEEN the Russians and Americans and told friends that nothing would drag him back into that field. But just when he thought he was out, two years later an opportunity he couldn't refuse pulled him back in. A Dutch company, founded by space advocate Rick Tumlinson and telecoms investor Walt Anderson, intended to lease the Mir space station from the Russians for commercial purposes, including space tourism and TV events. The deal was discouraged by NASA, which was advocating for the removal of Mir to focus on the ISS, and the activities planned for the station were again mocked when they were announced in 2000, according to Manber. The venture was doomed to fail, as Russia bowed to pressure from NASA to deorbit the space station in 2001, but not before Manber had put together some business proposals that were groundbreaking then, but familiar today.[9]

"During that time we signed with Dennis Tito to be the first self-paying tourist," he recalled. "We signed with Mark Burnett of *Survivor* to do a game show where the winner would go to space with the American television channel NBC. We signed with a number of different organizations and when the Mir was finally forced down, we had over $179 million in backlog, so it's a classic case where we proved the business model, that you can come into an aging space station at the end of its life, put some money into it, fix it up a little bit, get business. But there were too many political obstacles. And so the Mir was deorbited."

While the Mir was decommissioned, the ISS was on the ascent. But had things gone differently, there could easily have been a commercially run space station to rival the ISS. Today, a fully commercial space station is on the horizon (as we'll see in chapter 10), but still hasn't been realized. The presence of a privately run station orbiting the Earth would bring a new dimension to an off-world economy, and a boost to anyone looking to make money in space. The world would have to wait for that, and although the 1990s were ultimately a disappointment for the likes of Manber and Stone, who were advocating for a commercial space industry, they planted the seeds that would begin to bear fruit in the new millennium. Dennis Tito, nearly a client of Manber's on the Mir, was still determined to become the first person to pay for a trip into space, and companies that could make that a reality began to emerge. The commercial space industry was becoming more visible, and it became clear that the ultra-wealthy were ready to pay a fortune for a trip off the planet.

3

A $20 Million Ticket

ESTHER DYSON BEGAN INVESTING MONEY INTO FLEDGLING space companies in the early 2000s. In some ways it was the family business.[1] Her father is the famous physicist Freeman Dyson, who, among other achievements, worked on a nuclear-powered space rocket in the 1950s. Her mother was Verena Huber-Dyson, a brilliant mathematician best known for her work in group theory, an essential component of abstract algebra. Dyson herself has been a highly successful technology investor and a strong advocate for boundary-pushing science, including the quest to send humans ever further into space, and her ultimate goal is to retire on Mars.

But it hasn't always been an area of professional interest. When Dyson was a child, she knew that her father's generation had seen commercial airline flights go from a rarity to an everyday occurrence. "I kind of assumed the same thing—by the time I was grown up everybody would be flying to the Moon," Dyson told me from the New York office of one of her more recent investments, the tech startup Meetup. "So I didn't really honestly take that much interest in it. I mean, I sort of took it for granted and I thought the Moon was cool, but I was exactly not fascinated in it because I thought it had been taken care of. And then kind of around 2000, I began to notice, good grief, this space stuff. What's going on? It hasn't happened. So I started to get interested in it."

At that time, Dyson was running PC Forum, a globally revered technology conference, and created a more space-oriented event to complement it, Flight School, which ran for a few years before the 2008 recession hit. At this point, Dyson had already invested in two space companies: Zero-G and Space Adventures. Zero-G offers flights from commercial airports in the United States that give passengers the chance to experience, as the name suggests, zero gravity. The company was later acquired by Space Adventures, the only organization to date to have taken tourists to the International Space Station (ISS). "For me, as an investor, why invest in something that's going to happen anyway? The purpose of investing is to help stuff to happen," she said of those heady first days of commercial undertakings in space.

Dyson explained that the early era of space investment in the United States was seen as an affront to some members of the space and science community. "The internet was originally government funded and then suddenly everything changed and commercial interests came in. And somebody got into huge trouble, I think from Digital Equipment, because they sent an email message around the internet inviting people to a sales meeting and everybody thought that was disgusting. Same way you put a decal on a rocket, and people thought that was disgusting—[the idea of] commercial rockets."

Space Adventures would elicit a similar response when it began taking paying customers to space, starting with Tito in 2001. The American businessman had originally thought he'd be traveling to the Mir space station, in the deal involving Manber and MirCorp, but when that fell through and Mir was brought back to Earth, he turned to Space Adventures, who secured him passage on a Russian flight to the ISS, much to the chagrin of the Americans. A great deal of friction between Russia and the United States ensued, as NASA was unhappy with an untrained astronaut spending any time on the ISS. The Space Adventures president at the time, Tom Shelley, would later say that NASA did everything it could to prevent Tito making it to the ISS.[2]

Tito and the two Russian cosmonauts were sent to the Johnson Space Center in the United States to receive training for the American section of the space station. But when they got there, NASA's manager Robert Cabana refused to train Tito, according to Greg Klerkx in his 2004 book *Lost in Space: The Fall of NASA and the Dream of a New Space Age*.[3] One of Tito's Russian crewmates, Talgat Musabayev, told NASA that either the entire crew would be trained or none of them, to which Cabana replied: "In that case, we will not be able to begin training, because we are not willing to train with Dennis Tito." However, Tito had brought a *Newsweek* reporter along with him, and when the article ran the following week, NASA was made to look more like a bully, protecting an old way of life, than an agency dedicated to moving forward into the universe. Tito got his flight to the ISS, but with certain restrictions. He agreed to pay for anything he broke on the space station and signed an agreement that he would not sue anyone if he was injured. He was also instructed to sleep only in the Russian sections of the station. Even while he was onboard the international squabbling continued, as the American space agency demanded Russia pay for lost research time while Tito was aboard.

Tito's seven days in space were life altering. In February 2013, he set up the Inspiration Mars Foundation, an attempt to send a privately financed spaceflight to Mars by 2018, but admitted defeat on the project in 2013. He did inspire others to try out space tourism, however, and follow his lead as the first private citizen to pay their way into space. Not only did this legitimize the Space Adventures business, it also made Tito something of a star in the commercial space world. Klerkx described him as "the poster boy for a new era in human space travel: the man who had knocked down the gates for the rest of us who might like to someday visit the final frontier."[4] In reality, only six more people have visited space as tourists, each of them through Space Adventures, although there are plenty of companies gearing up to revive that business model.

Dyson herself trained to be an astronaut as part of the Space Adventures experience and floated in zero gravity. As part of her training, she underwent extensive medical examinations, which revealed she had Barrett's esophagus, a condition that could lead to a very deadly form of cancer. Armed with this knowledge, she underwent regular checkups and caught the disease early, despite it being extremely difficult to diagnose. She was able to treat it and in a 2017 interview said training for space saved her life.

*   *   *

WHILE TITO WAS BLASTING OFF TO THE ISS, ANOTHER WEALTHY man was plotting his own way off the planet, but his goals were decidedly more grand. As a young man in South Africa, Elon Musk had loved to read, and among his favorites was Isaac Asimov's *Foundation* series, in which the hero, Hari Seldon, invents a way to predict the future based on crowd behavior. Upon seeing a future dark age, he sends scientific colonies to other planets to protect against disaster. "The lesson I drew from that is you should try to take the set of actions that are likely to prolong civilization, minimize the probability of a dark age and reduce the length of a dark age if there is one," he later said in an interview.[5]

But Musk took a roundabout route toward saving the species. In Canada, he studied physics and economics before transferring in 1992 to the University of Pennsylvania. From there he moved to Stanford in California to take on a PhD in energy physics. However, he quit after two days to start Zip2, a platform for online newspapers, with his younger brother, Kimbal. They sold the company in 1999 to Compaq for $300 million and Elon pocketed $22 million. He used some of this windfall to start an online bank called X.com, which later became PayPal.

In 2002, PayPal was sold to eBay for $1.5 billion, and Musk, aged 31, earned roughly $180 million from the sale.

Prior to this windfall, Musk had been working on an idea to reach Mars, an ambition he is considerably closer to achieving today. As he explored the possibilities in front of him, he got in touch with an American group called the Mars Society, led by the charismatic and sometimes controversial Dr. Robert Zubrin. "Musk read my book *The Case for Mars* and that had made an impression on him, and as result of that he came to a Mars Society fundraiser in the San Jose area, Silicon Valley, in late spring 2001," recalled Zubrin in a phone conversation. Zubrin is not the easiest person to reach, and the topic of Musk is one of the few he's willing to entertain in detail. "He gave $5,000 then and $100,000 after and he joined the Mars Society Board."[6]

At that time, the Mars Society was working on a project in which it would send mice into Earth's orbit in a capsule that would spin to create a gravity effect similar to that on Mars. This would allow scientists to study the effects of Mars's gravity on adult mammals born and raised on Earth versus mammals who have known nothing aside from Mars's gravity. According to Zubrin, there is little or no data on how astronauts would fare in other gravities if they were born and brought up in that scenario, and what the health implications would be. Musk was interested in that project and increased his involvement in the Mars Society further. But he also had his own ambitions for sending plants and eventually humans to Mars, and he turned to Zubrin for advice.

"I connected him up with a fella named Jim Cantrell, a friend of mine, who was at that point an independent consultant, to be his advisor," says Zubrin. "And they went to Russia together and tried to find a launch vehicle and they ran into this whole crazy situation of con men over there. People asking you for $50,000 before they'll quote you a price and all this nonsense." Musk and his party encountered a number of strange dealings on their trip to Russia. One group reportedly insisted they make several toasts with vodka, until both Musk and Cantrell passed out. After agreeing to buy three rockets for $21 million, the Russians then changed their price to $21 million each, and when Musk complained, they responded: "Oh, little boy, you don't have the money?"[7]

When Musk returned, he started his own rocket company—SpaceX—which prompted what Zubrin describes as a "heart-to-heart discussion" with the Mars Society founder. Zubrin recalls Musk saying that buying Russian rockets wasn't going to work, and he preferred to lead his own operation rather than be reliant on others. He was also working out what to do with his future. One option was to open up the cosmos to humanity, and the other was to revolutionize solar energy here on Earth. "He agreed with the thesis of *The Case for Mars*, that Mars was the goal to make humanity a spacefaring species," said Zubrin. He essentially told Musk that the solar power revolution would go ahead with or without him, whereas a man of his wealth was needed to kick-start the mission to Mars. Musk listened to Zubrin and prioritized the Mars mission (though he started a solar power company and a car company at the same time).

SpaceX established its first headquarters in an old warehouse in the Los Angeles suburb of El Segundo. Musk's plan was to build as much of the machinery in-house, rather than buying it from suppliers all over the world. He wanted to ensure that SpaceX was faster and cheaper than his competitors in every way. The first rocket the company planned to build was called Falcon 1, after the Millennium Falcon from *Star Wars*. Musk wanted the rocket to carry a 1,400-pound payload into space for $6.9 million, an incredible feat considering that at this time a 550-pound payload would cost a minimum of $30 million to send off into space.[8] The company went about designing and testing engines, launchpads, and various other objects, and an astonishingly fast six years later would change the entire landscape of the space industry. Musk wasn't the first wealthy tech entrepreneur with spacefaring ambitions. Amazon founder Jeff Bezos founded his space company, Blue Origin, in 2000, although the existence of his startup was kept secret until 2003, when it bought land in Texas. Slowly but surely, a new breed of space companies was being formed, and in 2004, there was another great leap in the direction of commercial space.

*   *   *

STEPHEN ATTENBOROUGH WAS ON A CAREER HIATUS WHEN HE met Will Whitehorn for a drink at a bar in Sussex, England. It was a Friday evening in October 2004 when the two friends got together, and Attenborough had recently quit his job in banking.[9] Their discussion that evening would steer the course of Attenborough's life away from finance and into the business of spaceships, rockets, and holidays in space. Whitehorn worked for the Virgin Group founder Richard Branson, and had been tasked with keeping an eye on what was happening in commercial space circles. The British billionaire had already made a name for himself through various eccentric activities, including sailing the Atlantic Ocean and various hot air balloon feats, while his business ventures included the airline Virgin Atlantic. But ferrying passengers between continents was only the beginning of Branson's ambition. He wanted to take people to space. Branson, a *Star Trek* devotee, had been watching for any possible way he could make a successful entry into the space industry and make the show a reality. In 2004, he spotted an opportunity.

On that October evening, Whitehorn had just returned from the Mojave Desert, where he'd witnessed the culmination of the Ansari XPrize, a competition offering a $10 million prize to the first nongovernment organization to launch a reusable manned spacecraft into space twice within two weeks. In total, 26 teams participated, ranging from well-funded companies to enthusiastic hobbyists. The winning team was Tier One, run by a company called Scaled Composites—later found to be backed by Microsoft cofounder Paul Allen.[10] The competition was first announced in 1996 by Peter Diamandis, an engineer and entrepreneur who had already set up the International Space University and two attempts at private space companies that never really flourished. Through the XPrize, Diamandis sought to spur the private space flight industry in ways he couldn't as an entrepreneur.

Diamandis took inspiration from aviation contests of old, in particular the Orteig Prize—a $25,000 reward offered by the New York hotel owner Raymond Orteig to any aviator who could fly nonstop from New York to Paris or vice versa. The race was announced in 1919 and won eight years later by Charles Lindbergh in the Spirit of St. Louis aircraft. When Diamandis first announced the XPrize, there were no sponsors, no guarantee of the winnings, and initially no teams competing. By 2004, several teams were getting very close to achieving the goal, and Diamandis had to start thinking about the money.[11] In May that year, Anousheh and Amir Ansari donated a multimillion-dollar sum to the contest, which was subsequently renamed the Ansari XPrize. But the prize fund was still short, as Dyson remembers. "It was going to be a $10 million prize and he got about $3 million from the Ansaris and couldn't get to the $10 million," she recalled. "So he went and bought an insurance policy against anyone winning, which is just absolutely brilliant. You can't do that anymore. Now they see the risk of somebody winning it is actually quite high. So yeah, it took people of courage and slightly different attitudes to start."

The Tier One spacecraft was called SpaceShipOne, and it had an unorthodox launch procedure. The craft relied on a mother ship called White Knight, which would take off horizontally in the same manner as a plane, with the spacecraft attached. White Knight would slowly climb for an hour to an altitude of around 14 kilometers, before SpaceShipOne would disconnect. Once unattached, the spacecraft would glide, briefly unpowered, before a rocket ignition thrust it upward. By the end of the spacecraft's burn, it reached speeds of up to 3,000 feet per second and was capable of exceeding an altitude of 100 kilometers, outside the Earth's atmosphere. SpaceShipOne then reentered the atmosphere, decelerated rapidly, and eventually would glide back down to the ground, landing horizontally. On September 29, 2004, SpaceShipOne, piloted by Mike Melvill, reached an altitude of 102.9 kilometers, and on October 4 that year, secured the XPrize with a second flight piloted by Brian Binnie to 112 kilometers.

To Branson and Virgin, this was proof that commercial space flight was possible, and potentially profitable. So when White-horn met Attenborough for a drink, they got to talking about an interesting prospect. Attenborough, a polished talker with a prominent tan and gleaming set of white teeth, recalled the evening's conversation when he and I met at Virgin's offices in New York. "As we were talking, [Whitehorn] just said: 'Actually, you're not working at the moment, are you? We want to get this thing up and running and, you know, because we've announced early, we don't really have anybody particularly dedicated to this within the group. If you'd like to come along or if you're inter-ested in coming along and just helping us get a few of the sort of foundation stones in place, then why not?'"

Despite having no experience in space or any of the related sciences, Attenborough signed on as the first employee of Virgin Galactic, Branson's new company, which aimed to take the tech-nology showcased by the Ansari XPrize and use it to whisk tour-ists up into space for brief but memorable vacations. Branson had announced a partnership with Scaled Composites in September 2004, and the company was officially unveiled to the world. "We hope to create thousands of astronauts over the next few years and bring alive their dream of seeing the majestic beauty of our planet from above, the stars in all their glory and the amazing sensation of weightlessness," Branson said. "The development will also allow every country in the world to have their own astronauts rather than the privileged few."

The company slapped a $200,000 price tag on a trip to space, although it couldn't tell its customers exactly when they would be blasting off. "We set up a little website in 2004 and we launched the company and we played the SpaceShipOne video," Attenborough said. "We basically said, 'Look, there's a lot we don't know, but we think we have a really interesting opportunity here and if you'd be interested in reserving one of the early seats, then sign up here and we're also going to require a check for $200,000,' which . . . at the time we thought was the right sort of pricing level."

And plenty of people did show an interest from the start. The company didn't reveal the exact number of tickets sold, but Attenborough said they were "very successful very early." The next year, Virgin and Scaled Composites unveiled a joint venture called the Spaceship Company, which would manufacture SpaceShipTwo, the next generation of spacecraft and successor to the XPrize winner. This company would also build a new version of the mother ship, to be called WhiteKnightTwo, and was seemingly just years away from taking people to space for a fraction of the price Dennis Tito paid. As it turned out, the Spaceship Company took a little longer than just a few years, as the company was beset by technical difficulties, tragedies, and delays. But the would-be space travelers kept hold of their tickets and may be rewarded very soon. Today it is clear that the winning of the Ansari XPrize—and the formation of Virgin Galactic—was a landmark moment in commercial space.

The five-year period at the turn of the millennium saw many of the building blocks put in place that would eventually lead to a thriving private sector in space. The major wealthy backers got involved: Musk, Bezos, Allen, and Branson, and there was a proof of concept that people (at least very rich people) were willing to pay for services such as space tourism. But many still scoffed at the ambitions of entrepreneurs like Musk, and the treatment of Tito demonstrated the disdain with which NASA held certain parts of commercial space. Meanwhile, NASA was having major problems, and the possibility that the agency would have to ground the shuttle program—its most reliable form of transport to the ISS—opened up a host of possibilities for private companies to reverse their relationship with the American space agency. The shuttle had achieved great feats and contributed to the world's most ambitious project in space, but public perception was stained by its tragic failures, and a whole generation was left wondering what had happened to the NASA that took people to the Moon and back. Soon the agency would become the client of commercial space companies, hitching a lift on privately operated rockets.

4

NASA:
From Contractor to Client

IN FEBRUARY 2003, NASA'S SPACE SHUTTLE PROGRAM SUFFERED A second catastrophe, after the Challenger disaster of 1986. The Columbia spacecraft disintegrated on reentry, killing the seven people onboard.[1] The Shuttle program was grounded for two years, and NASA was forced to think of other ways the United States could ferry its astronauts and supplies to the International Space Station (ISS). The world was again reminded how dangerous space travel really is, and the Columbia tragedy led to renewed calls that the United States was better off spending its money elsewhere. The fate of America's future in space was deeply uncertain.

One potential solution came in 2005, when Congress put $500 million into a public-private partnerships program called Commercial Orbital Transportation Services (COTS). While a modest amount of money in the context of space flight, the program laid the groundwork for private companies to work with NASA. The goal of COTS, according to the agency, was "challenging private industry to establish capabilities and services that can open new space markets while meeting the logistics transportation needs of the International Space Station."[2] This program allowed NASA to shift the risk of sending cargo and people to space into the hands of commercial companies, something that appealed greatly after the Columbia disaster. Under this agreement, NASA wouldn't have to oversee the construction of pieces of a shuttle or rocket, and they wouldn't be paying out of their own funds for companies to build something the agency itself would operate. It was a much simpler and cheaper arrangement for an agency that had seen its budget cut at every opportunity since the end of the space race.

The $500 million investment represented less than 1 percent of NASA's total budget, but optimists still saw COTS as a backup to ensure that the existing and future plans of the agency to build and launch its own rockets to the ISS continued apace. Those plans had been announced dramatically by President George W. Bush and involved the Constellation program, which promised to do two things: send humans back to the Moon and ensure NASA still had access to space after the Space Shuttle program ended.[3] Bush also announced missions that would confirm the presence of ice (thus, water) in the Moon's craters, which would lead to the Resource Prospector, a NASA-led mission to mine ice from the lunar surface.

While the COTS program was introduced, Musk and SpaceX were still building their rocket, and by 2006 the Falcon 1 was ready for launch. The company was unable to secure a launch at any nearby test sites, so had to take the first iteration of the Falcon 1 to Kwajalein Island, known as the Kwaj, part of the Republic of the Marshall Islands. The rocket was hauled there via barge, and the SpaceX team traveled either on Musk's private jet or commercial flights. After a number of delays, they finally had the right conditions for launch on March 24. The rocket ignited and began to climb. Musk himself was watching as, after 25 seconds, a fire broke out above the engine and the rocket began to tumble back toward the ground. The Falcon 1 fell directly onto the launch site.

This was a huge blow to Musk and SpaceX's plans, but not a fatal one. Despite concerns over money, Musk pushed ahead with two more launches, both of which failed. On the fourth attempt, disaster struck even before the rocket reached the launch site. Musk and the team decided to fly the rocket to the Kwaj rather than send it by boat. But when the hired military cargo plane began to land in Hawaii, the pressure crumpled the body of the rocket. Engineers on board the plane managed to mitigate some of the damage, but the incident threatened to derail the launch by months. Musk sent an extra team of engineers to fix the rocket, and two weeks later the launch was back on. This time, the fourth attempt, on September 28, 2008, the launch was a complete success.[4]

Relief swept the SpaceX team—they'd just put the first privately built, liquid-fueled rocket into orbit. But as SpaceX reveled in its biggest success to date, Musk was battling a problem that could bring him back to the ground with a bang—he had nearly run out of money. He had invested $70 million of his own into Tesla, where he served as CEO. Keeping both Tesla and SpaceX afloat almost cost him his dream of sending rockets into space. Tesla, beset by production delays and fierce criticism from all sides, was at one point hemorrhaging $4 million a month, and by 2008 needed a major round of investment just to keep it alive. Musk used all the contacts he had to scrape together a tentative agreement from investors to put up the cash. However, one investor pulled out at the last minute, meaning Musk had to completely restructure the deal. Without investment, Tesla was on track to become bankrupt. Musk was faced with the decision of saving one of his companies or losing both. And then, when all hope for both companies appeared to be lost, on December 23, 2008, SpaceX won a $1.6 billion contract for 12 NASA resupply flights to the ISS, as part of the COTS program. Then, on Christmas Eve, the Tesla investment deal closed, and the electric car manufacturer was saved. Musk broke down in tears after both deals came through.[5]

The following year, in July 2009, the Falcon 1 became the first privately developed liquid-fuel rocket to deliver a commercial satellite into orbit. The era of commercial space flight had truly begun, and the balance of power shifted dramatically from the public to the private sector. When Barack Obama came to power, he declared Bush's Constellation program to be "over budget, behind schedule, and lacking in innovation."[6] It was canceled in 2010, which meant that America couldn't send its own astronauts up to the ISS on NASA spacecraft.

Politicians, particularly those in states that relied on NASA work for job creation, asked for a new program, and Obama obliged. When he signed the NASA Authorization Act of 2010 into law, he was confirming a bill directing NASA to create the Space Launch System (SLS), the world's most powerful rocket, capable of taking a 70-ton payload into the Earth's lower orbit, a figure that would increase to 130 tons following future upgrades.[7] Like Bush's, Obama's program aimed to take Americans back to the surface of the Moon, and eventually to Mars. The operational date for the first vehicle was set for December 31, 2016, and the initial budget laid out was a reported $7 billion. But the rocket is yet to launch, and the undertaking has already racked up a total of $11.9 billion in expenses.[8]

In October 2018, an audit of the project went some way toward explaining why. The report, which was carried out by NASA's Office of Inspector General and made available to the public, was fairly damning in its evaluation of how the project is behind by 2.5 years already and billions of dollars over budget.[9] Much of the blame was placed on one private contractor—Boeing. The aircraft manufacturing company, best known for producing the most widely used passenger airplanes in the world, was tasked to deliver some of the largest parts of the SLS, including the main component of the rocket, the core stage. The aerospace giant was told to deliver that in time for the original scheduled launch date of December 2017, but it hasn't yet been completed. The firm is also tasked with building other parts of the rocket, including the "exploration upper stage," which was expected to be delivered in mid-2021. Current estimates suggest that part won't be ready until about mid-2022. In a slightly bizarre video accompanying the report, a director within NASA's Office of Inspector General, Ridge Bowman, criticized one of the world's largest companies in a monotone voice with little or no drama. Despite the delivery, this was a major rebuke. "Our audit work determined that the cost increases and schedule delays can be traced largely to management, technical, and infrastructure issues driven by Boeing's poor performance," Bowman said. The report itself also blamed Boeing, but stated NASA had to take a share of the responsibility. "We believe a lack of action by senior leadership in both organizations to correct identified problems remains a significant cause of the SLS Program's cost increases and schedule delays," the report read. "Unless senior officials at NASA and Boeing are involved and collectively agree to the solutions, launch dates will continue to slip and the costs will increase, raising questions about the Program's long-term sustainability."

The report doesn't mention perhaps the SLS's greatest shortcoming—the rockets aren't even reusable. As SpaceX and Blue Origin have brought prices down with rockets that can be used multiple times, NASA and Boeing have seemingly been making more expensive rockets that are less recyclable. That's extra frustrating for the likes of Bill Stone, who wanted to make the space shuttle's external tanks reusable all those years ago, and who is now still planning a refueling point close to the Moon. "The problem with SLS is that Boeing/Lockheed are still living like it's the Cold War and enjoying the cost-plus contracts. To them, there is every motivation to milk that system as far into the future as possible. It's just two different ways of thinking about how to make money as a business. Ultimately, the Cold War funding model cannot survive—because it is a negative return on investment," Stone wrote in an email a couple of weeks after being interviewed. It's clear this is a topic that angers him, and for good reason.

Dallas Bienhoff was working at Boeing at the time of the project, and was with the company for 35 years, until May 2017. That was when he decided he was going to start his own venture, to do things that "were not possible within Boeing."[10] In 2004, Bienhoff wanted to put space tugs in orbit to move spacecraft around and a propellant transfer station in the Earth-Moon L1 location (a point in between the Earth and the Moon where an object could maintain its position between them, described best as an orbital parking lot). This was a similar idea to Stone's, in that it would put infrastructure in space to ease the burden of rockets and spacecraft launching beyond Earth's orbit. But when the subject was brought up within Boeing, it wasn't well received. "Boeing has the Space Launch System contract and therefore the things that I think need to be done are 'anti-contract,' if you will," said Bienhoff. "They were seen by many within NASA, and therefore within Boeing, as the enemy of the Space Launch System." He says with complete conviction that this attitude set back plans to put reusable and cost-efficient infrastructure in space. "From a financial bottom line standpoint for the company, you don't want to throw away [a] $4-billion-a-year contract, but from an innovative standpoint it hurts," he said. "Once Boeing was awarded the SLS contract, which was late 2010, I was not allowed to talk about depots and the same was true for my counterparts at ULA [United Launch Alliance] because Boeing is half owner of ULA. And so we both had these throttles put on us, and I think it stymied innovation and stymied progress in how to effectively get to the Moon and Mars." Bienhoff has now set up his own project, the Cislunar Space Development Company, which promotes the idea of space tugs, fuel depots in space, and reusable infrastructure in general. Bienhoff has no desire to build the hardware himself, but sees his company as a private version of NASA, which can create the plans and work with manufacturers to make it happen.

The new wave of private companies like SpaceX and Blue Origin have been competing with the SLS program since it was established, and they are unafraid to take bold steps forward. Their reusable rockets are cheaper, more efficient, and actually work. Around the same time the SLS was announced, Boeing was scooping up more government contracts, this time from the military, through a joint venture with fellow rocket-builder Lockheed Martin. Once again, Musk was the man to challenge the status quo, claiming SpaceX could do the same job for cheaper, and he came out victorious once more.

* * *

THE U.S. MILITARY HAS LONG FELT IT VITAL TO MAINTAIN steady access to space. The nature of the technology it launches is often classified, but it has routinely catapulted spying, navigation, and other satellites into space since the military's first launch of a reconnaissance satellite in 1959. Initially, this national security infrastructure was sent into space on modified intercontinental ballistic missiles, but by the 1970s the U.S. Air Force had turned to NASA for help. The Air Force worked with the space shuttle program to establish a means to send more defense satellites into space, and in June 1982, the shuttle took its first Department of Defense payload into orbit. But the Challenger accident of 1986 changed that, as it changed so many aspects of America's space program, and the Reagan administration ordered the military to develop a "mixed fleet" policy for launching its hardware.[11] The last shuttle launch associated with the military was at the end of 1992.

The military turned to commercial rocket launchers, and two companies in particular: Boeing and Lockheed Martin. Lockheed had its Atlas rockets, while Boeing offered the more expensive Delta. They both wanted to capture a significant slice of the commercial satellite market, but their rockets were too expensive to compete with models from Russia and the European Space Agency. So they both bet big, in hopes of clinching billions of dollars worth of military contracts. This led to some fairly shady practices during the early 2000s. The U.S. Department of Justice began an investigation into how Boeing had got its hands on tens of thousands of pages of trade secrets from its competitor Lockheed Martin and lawsuits ensued.[12] At this point, the U.S. military became concerned its supply of rockets to space could be hindered, and in 2005 it stepped in to broker a deal between the two companies and merge their rocket-building departments into one organization—the United Launch Alliance (ULA). Boeing and Lockheed Martin each had a 50 percent stake in the new company, and the ULA had a monopoly on military space contracts in the United States.[13]

According to Barry Lynn, the executive director at the Open Markets Institute, the real danger with monopolies arises when there's the dual threat of no competition and poor regulation. "The more competition you have, the more incentive [there is for] the people who are competing in the space to keep up their standards," he said.[14] "The most dangerous situation is when you have a private monopoly that's not regulated by competition and not well regulated by any kind of non-corrupt agency." ULA was still being well regulated by NASA and other agencies for its launches, which mitigated some of the problems, but the lack of competition still riled many people. One company stood up and objected. Musk's SpaceX was still in its infancy in 2005, but it too had its eye on the billions of dollars to be made from military contracts. Musk sued to stop the merger in October that year, claiming the proposed merger created a monopoly.[15] But SpaceX, still a year away from its first attempt at launching a rocket, lacked the influence to force the issue. The ULA won antitrust clearance from the Federal Communications Commission in 2006, and the merger went through. If the ULA and the military thought they'd heard the last from Musk, they were very much mistaken.

Between 2006 and 2013, two different contracts were agreed upon between the federal government and ULA. The first was a "firm fixed price" award, which was for a specific, individual launch. The second was much more controversial, an annual launch capability contract that involved the military paying a fee to ULA every year to be ready to launch whatever the government wanted. According to reporting by *Ars Technica*, this fee was nearly $1 billion per year and was paid regardless of how many rockets were actually used.[16] ULA wasn't just sitting back and collecting its money though. To give the military the guaranteed year-round access to space it craved, ULA maintained five different launch pads and a fleet of all different types of rockets. And it delivered impressively: ULA had a perfect launch record during this period, never losing a payload. In 2014, Musk decided it was about time SpaceX, which by this point was launching its Falcon 9 rocket consistently, got some of the action. But the military and ULA had other ideas. In January of that year, the Air Force announced an agreement with ULA for 36 national security launches by 2020. ULA claimed this bulk purchase of flights would save the government $4.4 billon. Musk saw it as yet more evidence of a monopoly and that his company had been frozen out again. The Air Force argued that SpaceX wasn't allowed to bid for any of these missions as it didn't have the certification required to fly national security missions, according to Eric Berger's 2017 reporting of the events for *Ars Technica*.[17] Musk was apparently furious, and in April 2014 was invited to a U.S. Senate Appropriations Committee hearing on national security launch programs, along with ULA CEO Mike Gass.

"It's interesting to note that from the time when Lockheed Martin and Boeing's businesses merged, the point at which they stopped being competitors, the cost has doubled since then," Musk said at the hearing.[18] He also brought up the annual launch capability payments, claiming they were a government subsidy that SpaceX wasn't receiving, giving ULA an unfair advantage. Gass quickly refuted the claim they were any such thing. Other voices spoke out against Musk's argument, with some claiming the issue of competition and a free market was moot when it pertained to national security issues, and the U.S. military relied heavily on the launches to provide crucial defense initiatives like surveillance and global positioning. Still more arguments pointed to ULA's spotless record as proof that the system worked. "I agree that competition typically results in better quality and lower price contracts. But the launch market is not typical," Republican senator Richard Shelby said at the hearing. "While the goal of competition is to lower the cost of access to space—which I think is good—combined with a need to maintain performance and reliability, such as we have today, competition may not actually result in a price reduction for the federal government."[19]

Musk's claims that costs had risen appeared to be true. By 2015, the U.S. government was paying an average of $225 million per launch, according to ULA. The Government Accountability Office put the figure at $360 million per launch, but did note that not all of that figure goes to ULA. SpaceX was offering to launch for just $100 million. "ULA has decided that they're afraid even of an unfair competition," Musk told *Bloomberg* in 2015.[20] "They don't want a fair competition. They don't even want an unfair competition. They want no competition at all."

Two weeks after the hearing, Musk filed a lawsuit against the government, seeking the right to bid for national security missions and protesting the package launch bid process that had secured ULA launches up until 2020. This forced the government to the negotiating table, and in an out-of-court settlement, the military agreed to certify SpaceX for national security launches in May 2015.

Musk's litigation and noisy dissent worked, and the Air Force also opened up some of the launches to competition. SpaceX was awarded its first contract from the Air Force in April 2016, worth $82.7 million, and launched its first rocket with a national security payload in May 2017.[21] By the end of 2018, the company had secured seven launch contracts from the military: five of which are for the Falcon 9 and two for the Falcon Heavy, but even as the military moves on from the ULA monopoly, the criticism continues. "We did a really, really bad job of managing the competition back when they gave the go-ahead to allow Boeing and Lockheed to merge their launch systems," said Lynn. "That was really stupid. I mean it was either stupid or corrupt, but from the point of view of the American people it was bad. That was really, really bad regulation."

Toward the end of the year, the Air Force started to expand the number of companies considered for future launch contracts. The military branch gave out three contracts worth a total of $2 billion to Jeff Bezos's Blue Origin, Northrop Grumman Innovation Systems, and ULA to develop new prototypes for launch systems. The money is given to the companies to work on new rockets, and each company received $181 million. Through 2024, $500 million will be given to Blue Origin to develop its New Glenn rocket, $792 million to Northrop Grumman for the OmegA, and $967 million for ULA's Vulcan Centaur rockets.[22] The deals are known as launch service agreements (LSA) and are meant to ensure that the Air Force has plenty of options to choose from for its launches. It's a long way from the days of the ULA monopoly, but there is one notable name missing from the list—SpaceX. Musk's company has benefitted from an LSA in the past, but wasn't awarded anything on the most recent round of contracts. The Air Force didn't confirm if SpaceX had even bid for the agreement, but did say the decision wouldn't stop the company bidding for future launches. Musk asked for increased competition, and it seems he will have plenty in the years to come. Whether any of the new rockets can beat SpaceX for cost and efficiency remains to be seen.

*   *   *

THE NATIONAL SPACE AGENCIES OF THE WORLD HAVE NOW SEEN the benefits of becoming customers of private companies rather than their employers. They can save hundreds and millions of dollars, and crucially can pass much of the risk on to the private companies themselves. According to many in the industry, this risk aversion has always held NASA back. "It all stems from the whole Cold War beginnings of the space program, where national heroes rode on rockets to combat the evil other nation," Stone said.[23] "And the problem then is if you lose a national hero, you look bad. And so the result was they tried to reduce risk of losing a national hero to zero and you can't do that. You can't economically do that and stay in business, which is why, for example, you see things like the space station taking 30 years to complete. You see the SLS launch system taking 20 years, and they still haven't launched a rocket."

NASA has now effectively ceded a large part of the space industry to the private sector, while it concentrates on exploring deeper space and other planets and moons, for example, putting rovers on Mars and sending probes to the likes of Saturn. But Bienhoff believes this is only the start for private space companies. "The things that we, the space community—the moon huggers, the Mars huggers—want to see done far outstrip the budget that's available within NASA and the rest of the world combined," he said. "There has to be money made by the people doing services or building products that others will buy of their own volition so that they can do more."

The age of NASA employing private companies to work on government-funded projects in space was over, and the agency had settled for a place as a customer and enabler of industry. Capitalism and space now mix more than they ever have before, and for Bill Stone, it couldn't come soon enough. "I'm talking about private enterprise pulling off what government cannot do," said Stone. "They are not structured to do it. The government isn't designed to make a profit, not unless you're an Attila the Hun and you raid a country, kill all its people, and take its women and gold. That's how you make a profit if you're talking about imperialism, but that's not what people are into these days." The threat of space imperialism may largely be over, but

now we must face the consequences of hundreds of self-interested startups flung out into the universe.

And there will be consequences. Capitalism has been a driver
of innovation here on Earth, but it has also caused rampant
inequality, lax regulations, and environmental disaster. This same
force now drives us toward space, and it won't be long before we
see the effects on the ground.

PART II

THE PRESENT

5

Money and Politics

WHEN ELON MUSK'S SPACEX SUCCESSFULLY LAUNCHED ITS FIRST rocket in 2009, the company did what no other rocket provider had done before—it made its prices transparent. This one very basic act enabled a whole host of companies to adjust their ambitions, build solid business plans, and solicit investments with real figures in hand. A new generation of space entrepreneurs were taken much more seriously, and the dollars started rolling in from private investors, banks, and the lifeblood of Silicon Valley's technology industry—venture capital firms.

In 2012, Chad Anderson wanted to find a place for himself in the space industry, despite having no expertise in physics or any of the sciences. In his own words, he is a "business guy, through and through."[1] At the time he was managing a commercial real estate portfolio with JPMorgan bank in Seattle, and although it paid well, he wasn't finding the job particularly rewarding. "I was looking for something more meaningful to do with myself and my career, and so I was doing a lot of stuff on the side," he explained from his offices in the financial district in New York. "I started a nonprofit focused on entrepreneurship on the whole [Middle East and North Africa] region, based in Morocco. I was doing a lot of work with the Seattle Children's Theatre. I was doing things that mattered to me." To move further in this direction, Anderson, a confident, likable, and precise speaker, moved to England to pursue an MBA at Oxford University. He chose Oxford because, as well as being one of the most prestigious universities in the world, it had a well-respected program for social entrepreneurship.

Anderson thinks space is the ultimate social venture because he believes that by solving problems in space, we can change the lives of more people on Earth. He points to issues like internet connectivity and climate research that are only possible with technology deployed outside the planet. He also believes by putting more people in space, people will come to appreciate the Earth more. That was the more abstract way in which he was drawn to space, but there were also practical reasons why the industry was attractive. "Space is really important and it impacts our world hugely," Anderson said. "If you get into how satellites affect our daily lives and how the world would not operate without GPS, how it runs our banking systems, like if you think about it, it's a critical global infrastructure. Our entire global economy is based on a backbone in space. And then you think about also how through space we're going to connect all these rural areas and these other 3 billion people in the world who don't have access to the internet, you know that when people connect to the internet their livelihoods improve, they have access to more services."

Upon graduating, Anderson started interviewing with people and companies in the industry to find a place for a man with his skillset and quickly perceived that he was in demand. The private space industry was booming, and it was full of technical people but not enough with a sense for business. At this point, Anderson started to see "different threads coming together," as SpaceX delivered its first cargo to the International Space Station (ISS) and the industry began to take off in earnest. "I've always relished in chaos and I prefer unstructured versus structured. I've always done really well there. And so for me, entrepreneurship comes naturally and I have a finance background so it was all these new companies that we're trying to get up and running," says Anderson. "What do they need? They need funding. So getting involved back then it was total chaos. Anderson recalls people making wild claims about space and not really evaluating the business side of their ideas properly. He put himself in a position where he could translate the risks and opportunities in space to the financial world and help investors understand what was really going on in the industry.

Anderson is now CEO of Space Angels, a prominent investor in space startups, operating out of New York City. The firm has invested in many of the more successful young space companies and also publishes data on the state of the industry. That data shows the space sector has come alive, something that Anderson attributes to the efforts of SpaceX. "In 2009, when SpaceX had their first successful commercial launch, they really brought down the barriers to entry. They brought the cost way down, but then they also published their pricing. And this transparency was unheard of in the market," Anderson said. "The defense contractors had made a living on the opaque market, the sort of cartel that they'd been running. And when SpaceX said we can take this much weight on Falcon 9 to this orbit and it costs $60 million, it changed everything. The barriers to entry come crashing down and in rushes this wave of entrepreneurship and innovation like we've seen in countless other industries. You can scroll through the business school case studies, there's tons of them that look just like this of an industry being disrupted."

Space Angels is thought to be the first space-focused investment firm in the world and allows entrepreneurs to bypass some of the roadblocks they face when looking to raise money for their companies. In the past, investors have worried that space companies are cash-hungry, long-term gambles loaded with risk. To address these concerns, Anderson and Space Angels have compiled the largest database of private investment transactions in space, covering pretty much every nongovernment equity investment in the space industry. They studied the amounts raised by space companies at the different stages of investment—seed, series A, series B, and so forth—and compared it to data for venture capital investments for every industry. The firm found that there was little difference in the amounts raised, going against the perceived wisdom that all space companies need shocking amounts of money to get them off the ground. "Space is broad and people think rockets and launch, right? Building a rocket company from scratch is expensive," said Anderson. "That is much more so than the average venture deal. But if you look at space broadly, it covers a lot. Launch is only one small piece of it. Satellite hardware is just another piece of it. Satellite platform data and analytics is another piece of it. There's a lot that isn't capital intensive. There [are] some that [are], but on the whole you look at the average of space versus general venture and it's pretty on par." Research by Space Angels finds that generally space companies begin generating revenue earlier than other venture capital investments, making them stronger businesses in the early stages in particular. "In an economic downturn, I would much rather be holding space company investments than general venture investments that are focused on getting users on their platform," Anderson said.

Since 2009, venture capital funds have invested $4.2 billion into space companies, with 70 percent of that amount coming in just the last three years.[2] As of the end of 2018, there are 534 venture capital firms with investments in space, including 41 percent of the top 100 venture funds. The majority of the top venture capital firms come from one place, Silicon Valley—the Northern California epicenter for technology startups, and a place with a surprisingly long history in the space business.

*   *   *

THE GOAL OF THE APOLLO PROGRAM WAS TO FULFILL THE challenge laid down by U.S. president John F. Kennedy in 1961, to put Americans on the Moon and return them safely to Earth.

Only massive technological breakthroughs could achieve this in such a short time, and it may be that the largest consequence of America's quest to reach Earth's lunar neighbor was to inspire and enable the technology industry we recognize today.

In order to land humans on the Moon, the lunar module— the part of the spacecraft that would take the astronauts down to the surface—needed a computer. Back then, as now, every ounce of weight on the spacecraft added thousands of dollars to the mission cost, so the module had to be just large enough to carry two astronauts, while the other member of the crew stayed with the main ship orbiting the Moon. The limiting factor was that in the 1960s computers were gigantic machines requiring huge amounts of power to do the simplest of tasks. This had to change, and the work done to shrink the computer for the Apollo mission led directly to the smartphones, laptops, and tablets used all over the world today. This was the jump-start needed for the consumer electronics industry to be born, and it helped establish Silicon Valley as the major global force in technology.

NASA scientists working on the Apollo missions knew one tragic accident could set the space program back decades. But the agency took at least one major gamble, according to space historian Paul Ceruzzi. Ceruzzi notes that in most elements of the mission, NASA designed backup systems to kick in should the primary technology fail.[3] But the Apollo guidance and navigation system was controlled by a single computer in both the command and lunar modules. There were no backup computers on board, as there wasn't enough space for them. In full knowledge that the computers absolutely had to work, NASA chose to use a completely new technology to build them: the integrated circuit, better known today as the silicon chip. The computers were being designed by MIT's Instrumentation Laboratory, which worked closely with Fairchild Semiconductor, the company that invented the chip and is now recognized as one of the most important companies in Silicon Valley history.

In many ways the story of Fairchild Semiconductor's founding has come to epitomize the whole Silicon Valley phenomenon. Early in the 1900s, a series of successful electronics companies had prospered while based on the southern San Francisco Peninsula, creating a base of technical expertise in the area. Drawn by this talent and the proximity to the respected Stanford University, a man named William Shockley moved to the Bay Area in 1956 and set up his Shockley Semiconductor Laboratory. That same year, Shockley won a Nobel Prize for his work bringing silicon semiconductors to the world. The next year was less successful, as eight of his senior employees, now known as "The Traitorous Eight," quit to form a company of their own. The employees were tired of Shockley's abrasive management style and were keen to free themselves from his leadership. By all accounts, Shockley was a deeply unpleasant man to work for, and in general.[4] In later life he wrote a paper arguing people with lower IQs should be paid to undergo voluntary sterilization, and when he died in 1989, aged 79, he was estranged from almost his entire family, including his children.

The Traitorous Eight went on to form Fairchild Semiconductor, where they landed a contract with NASA for the Apollo missions, but not before some vigorous testing to ensure the chips wouldn't fail. Fairchild offered MIT a whole line of chips that could be used to make the computer for the mission, but MIT chose to focus on one specific type of chip to ensure that Fairchild would gain experience in building it correctly. The chips were chosen for the Apollo manned missions, and despite the risks and unknowns involved, no Apollo Guidance Computer ever experienced a hardware failure while in space.

Now the space industry has come full circle, as Silicon Valley startups are making big money in space. While Anderson's Space Angels has been considered something of a trailblazer, more companies have begun to emerge with the goal of connecting space companies to seed money. Sidney Nakahodo set up the New York Space Alliance in 2016 to help startups find the investment they need to prosper. "The best way for us to make a contribution to the ecosystem is to facilitate or to bridge the gap between startups and resources that they wouldn't have access to otherwise," Nakahodo explained in a hotel lobby in New York, where he's based out of a coworking office.[5] "So sometimes for startups, it's pretty hard, for instance, to connect with investors, so investors are interested in hearing pitches, quite often they come to us at an earlier stage and the way that we help them, it's not only about making introductions but also preparing them to pitch to potentially interested parties." Occasionally that can mean being careful about how they present themselves. "What we notice is that we have to be sort of cautious in the way that we introduce an idea. If it's a [venture capital firm] that has invested in space, it's fine for us to describe a company as a space startup. But there's some prejudice against the idea of doing things in space because people associate space with risk, long-term, and bureaucracy."

The New York Space Alliance came together slowly over time. The organization was just a Meetup group originally, a few commercial space sector enthusiasts gathering to talk about the future. But a request for pitches from the International Space University in France attracted the group's attention. The university was looking for an American organization to host its think tank on commercial space and space entrepreneurship. Nakahodo jumped into action and put together a proposal that would connect the university with academic organizations in New York. The pitch was ultimately unsuccessful, but the group did reach the top two finalists, and enough momentum was generated to form the New York Space Alliance. In late 2018, the organization paired with NASA to facilitate the two-way relationship between the space agency and entrepreneurs. "Everyone has an interest in working with NASA either because they want to provide technology to NASA or because they want to partner with NASA in order to use their expertise," said Nakahodo. "So what we did was to create a program to support start-ups that want to work with NASA. And we signed a space act agreement with them in October."

The program with NASA has two components: one aspect gives startups access to NASA's resources, while the other helps technology flow from the agency to the private sector. "Right now anyone in the world can apply for NASA technology, to license, it. But of course we know that it's very hard because of bureaucracy, it takes time for them. So the program that we have designed, the collaboration with NASA, was how can we speed up this process." The upside for NASA is that the New York Space Alliance is able to broker deals with the best startups and filter those with less potential. Companies that really impress the national space agency get the first three years of license fees waived and access to resources such as people and facilities. According to Nakahodo, NASA wants its technology to create a social or economic impact, so in none of these deals do they take equity in a company. But for venture capitalists, equity is what they're buying with their investments, and indeed plenty of shares in hundreds of space companies are being snatched up by the venture world. Entrepreneurs are being handed large amounts of money to pursue their ambitions in space. Not yet on the scale to complete something like Bill Stone's hydrogen mining project, but significant checks are being written.

But venture capital firms aren't investing in space compa-nies just to see humanity make inroads into the cosmos—they want to see returns. That means an exit, the word venture capi-tal firms use to refer to a company being sold or taken public, the moment they have the chance to recoup their investment and make a profit. So far we haven't seen many major exits in the space industry, although Northrup Grumman's $7.8 billion purchase of launch company Orbital ATK in 2018 was signifi-cant. It makes sense for these companies to stay private rather than to trade on public markets, so they can focus on their mis-sion rather than attempt to appease shareholders. But Anderson imagines there will always be potential for major acquisitions by the bigger companies. "Things really started to pick up in 2012 and in the average VC cycle, you invest hoping to get a return in five to seven years from a company," he said. "We're right at that now." Anderson has seen major returns on investment lately, mainly for satellite companies, as they were the early movers. Over the next few years he predicts more and more small launch companies will join the market, and the established players will seek to buy them.

The influx of money from Silicon Valley venture capitalists will only increase as the industry matures, but will they have the know-how and experience to know when a space company is legitimate? "A lot of what happens in venture and a lot of what happens in San Francisco and the venture world generally is a lot of hype and there's a lot of momentum and fear of missing out," said Anderson, who added that this can lead venture capital firms to make investments they may come to regret. "At a certain point the rubber meets the road and the physics has to close, right? There's a minimum bar that you have to meet. I see deals getting done, like the physics just don't make sense. And so I'm confused as to how those things happen, what level of diligence they're going into and who they're consulting."

* * *

VENTURE CAPITAL GIVES COMPANIES THE FUNDS THEY NEED to achieve their goals, but it also puts pressure on them to get results. This is a big shift from the days when space projects were sponsored by academic grants and government contracts. It's important not to forget that the recipients of those government contracts, the massive aerospace companies and the newcomers, all rely on government spending for their income. And because it can never be untangled from terrestrial politics, space remains in the firm grasp of the governments of the world, but the new wave of space entrepreneurs are beginning to spend big money to lure politicians to their side.

During the 2010 election cycle, the SpaceX PAC, a political action committee that pools money from contributors and funnels it to political candidates, donated $67,900, with $4 going to Democrats for every $1 going to Republicans. In the same election cycle, Boeing's PAC made contributions worth more than $2.2 million.[6] At the time, Elon Musk was keen to point out how disadvantaged he was in Washington, D.C., with the megaspending of the major players in the aerospace industry. "We don't have nearly the political base as other companies do— at this time," he said in an interview. "They talk about 'super PACs.' We're a micro PAC. We may be 1 percent of their political power. So hopefully, we'll start with bringing it up to 2 percent. We have more will. And we will reach out to the public directly."

Today, SpaceX has increased spending on lobbying. In 2017, the company shelled out $1.96 million, according to the Center for Responsive Politics, and Musk was set to spend even more in 2018. His company's budget for political influence still doesn't compare to the likes of Boeing, which spent more than $16 million in Washington, D.C., the same year. But SpaceX has since won major contracts with NASA, and Boeing has reportedly gone to great lengths to halt Musk's progress, devoting money and effort to call the company's practices into question.

In July 2018, an op-ed column titled "Next Space Explorers Must Go Boldly—and Safely" appeared in a Memphis newspaper. The article criticized SpaceX for the potentially dangerous practice of allowing its crew to board the spacecraft before fueling the Falcon 9 rocket. Over the next few months, copies of the opinion piece also ran in the *Houston Chronicle*, a number of Alabama newspapers, *Albuquerque Journal*, *Florida Today*, and *the Washington Times*. *Ars Technica* reporter Eric Berger investigated and found evidence suggesting a Washington, D.C., PR firm that lists Boeing among its clients was behind placing the op-eds.[7] Each of the articles had the same byline, "retired spacecraft operator" Richard Hagar, who used to work for NASA and now lives in Tennessee. Berger reached out to the author and found that he hadn't actually submitted many of the articles. At least four of the six op-eds had come from two people using gmail.com addresses, and the names associated with them both worked for Law Media Group. The company described itself on its website as a 15-year-old firm that "develops and executes public, Hill, and agency-facing issue advocacy campaigns that shift the narrative in a changing world." Which sounds a lot like business-speak for lobbying. SourceWatch, a publication of the Center for Media and Democracy, which aims to provide information about corporate public relations campaigns, described the Law Media Group as a "secretive Washington D.C. public affairs firm" with a history of masking the financial sponsors behind the op-eds it has placed in the media.

Later on, *Business Insider* reporter Dave Mosher found further proof of Boeing's involvement.[8] Not only did Mosher discover the op-ed author used to work for a Boeing-affiliated company and described himself as a Boeing retiree, he also revealed that he only shared the article with one person—a Boeing employee he refused to name. Neither Boeing nor the Law Media Group commented to any media on the matter, and it's not been confirmed that either were involved, but the evidence is quite damning that Boeing went to extreme lengths to discredit a competitor.

This type of tactic is to be expected in an industry that increasingly sends more and more advocacy groups to Washington, D.C. The lobbying groups can be roughly sorted into three types. Some lobbyists represent a particular company, like SpaceX, and use any means in their power to exert influence to aid their client. Others represent an industry group, such as the launch companies like the United Launch Alliance. And the final group is perhaps the most likable. They lobby Washington to do all in its power to carry on a specific mission that a given group holds most dear—like the Explore Mars group.

Chris Carberry helped found Explore Mars in 2010, with the goal of advocating for policy that will take humans nearer to visiting the Red Planet. "One of the things we do best is working from inside out rather than outside in," Carberry said.[9] "So many advocacy groups basically are screaming at power and hoping they'll listen. Whereas what we do is we work well with people in the industry and kind of bring people together more subtly and quietly, work with them to find ways to move forward and build the community." The organization holds events to unite major players from academia, NASA, international space representatives, and, of course, the commercial sector.

Carberry has been lobbying U.S. Congress about space issues since the 1990s, and has seen plenty of change over that time. When he first visited congressional offices in 1997 or 1998, there was barely any support for missions sending humans to Mars. "It was considered science fiction and basically you'd go in, say, oh, we want to send humans to Mars, and you'd see eyes roll and it certainly had a giggle factor. Most, most members of Congress did not take you seriously at the time," he recalled. That has changed drastically in recent years, according to Carberry. He now finds it rare not to hear "extreme enthusiasm" and strong bipartisan support for human space exploration missions in general. "I can't actually remember a time ever when there's been such strong support, and that's a good thing and we should be able to move faster now and have more decisive policies based on that. But it's still been elusive despite that unprecedented support we have."

The sheer number of lobbyists from the space industry knocking on the doors of politicians in Washington, D.C., has definitely stoked this increased support, and Carberry has watched their ranks grow considerably over the last 10 years. "You have a lot of nonprofits that have been around for a while, but have become more politically active and go up to the Hill," explained Carberry. "And, of course, since there are a lot more commercial players, they all go up to the Hill with their lobbyists or whatever other methods they are using to gain political influence. There are a lot of people going up to the Hill, it's kind of funny, even that's changed. Because early on when I was looking at this, the space community was going to the Hill, but it was not as frequently at it should be. Now in any given time period we're tripping over each other. And I'm regularly going to Congressional Offices, and they say, 'Oh, another space group!' The messaging may not be exactly the same and some may be supporting the Moon, some may be supporting Mars. But for the average member of Congress, it doesn't really matter. The message all blends together and they say support for space, and . . . in the end, that's good enough for us, that Congress understands there is solid support from these various constituencies."

Lobbying for the space industry has already had a major impact on U.S. politics, and the president. In March 2018, President Donald Trump was visiting the Miramar Air Station in San Diego. While talking to a group of Marines, he suggested a plan as if it had just jumped into his head. "You know, I was saying it the other day, because we're doing a tremendous amount of work in space—I said maybe we need a new force. We'll call it the 'space force,'" he said.[10] "And I was not really serious. And then I said what a great idea—maybe we'll have to do that." Space Force is Trump's grand plan to make another division of the armed forces, one that would have the sole task of protecting America's interests in space. The details of the proposed military branch are still hazy, but the idea appears to be centered around keeping space superiority in the hands of the Americans. But the idea didn't originate from a moment of inspiration by America's president, it's been around for a while, and it stands to benefit greatly the commercial aerospace firms that have been lobbying hard for it for years.

Space Force is a proposed sixth branch of the U.S. military, after the Army, Navy, Air Force, Marine Corps, and Coast Guard. Currently, much of the military's dealings in space fall under the purview of the Air Force. Trump's vice president Mike Pence has said the administration will send a legislative proposal to create Space Force in 2019, and aims to have the program established by 2020.

According to reporting by the *LA Times* in August 2018, Space Force has been the pet project of a small group of current and former government officials, some of whom have deep financial ties to the aerospace industry, since 2016. These officials, and presumably the aerospace companies behind them, believe the creation of a new part of the military specifically for space will increase spending on satellites and other space systems. One of the biggest proponents of Space Force has been Representative Jim Cooper, a Democrat from Tennessee. But Cooper spoke out against Trump's "hijacking" of the Space Force idea, saying the president's announcement runs the risk of inflating the budget, as potential vendors get excited at the prospect of more government contracts. He claims that if private companies know Space Force is coming, they can adjust their prices accordingly and hit the government with a larger bill than before. But Cooper's biggest donors may already have known regardless. According to Open Secrets, which tracks money in Washington, D.C., Cooper received $38,500 from the aerospace industry during the 2018 election cycle, including $10,000 each from Boeing and Lockheed Martin.[11]

Space Force plans have largely been led by Mike Pence, the vice president who has taken a significant interest in the commercial space sector in particular. "We have been very fortunate in the space industry—I'll caveat by saying, unlike other industries—that there's a tremendous interest from the current administration and it's been a positive interest that we've seen," Eric Stallmer, president of the Commercial Spaceflight Federation (CSF), an industry association of businesses and organizations seeking to advance commercial spaceflight, told me.[12] "Vice President Pence is very much engaged and involved in the space industry."

The CSF provides different services to different people. For a company like SpaceX or Blue Origin, which already have plenty of lobbyists and employees in Washington, D.C., the organization is a unified voice speaking for the whole industry. The organization wouldn't go to Congress and ask for development money for an engine that these companies or other industry giants might be building. But for smaller space flight companies the CSF plays a vital advisory role and represents their voices in an environment that is more alien and unwelcome than space itself.

Stallmer joined the CSF in 2014, after 13 years running government relations for a commercial software company. A year after he began, the commercial space industry celebrated a significant achievement, as the Commercial Space Launch Competitiveness Act of 2015 was signed into law. This was the law that essentially made it legal to take resources from space without violating the original 1967 Outer Space Treaty. The treaty is notable for its ambiguous language, and this allows governments to interpret it in a number of ways. For example, the treaty says no country can claim any land on another planet or moon, but that doesn't mention taking resources. The United States essentially deemed it possible to exploit resources without claiming ownership of the land. Stallmer emphasizes the hard work that went into educating members of Congress that "this is needed, this industry is real and you can be a catalyst and you can be very helpful in moving this industry along, um, with, with sensible legislation that does not try to stifle the industry." In the end, he says the bill passed through the House and the Senate with bipartisan support. "You always say that when the bill is signed, but it didn't feel that way at the time," he added with a laugh. "But the legislation passed and we worked it every step of the way until it got to Obama's desk."

That bill was the first of many political victories Stallmer has helped win for the space industry. But when Donald Trump was elected president, he must have feared the worst, as many people did. Trump is notoriously anti-science. He is a climate change denier, and as his tweets regularly show he has very little interest in anything that isn't self-serving or appealing to his base. On the campaign trail in 2016, a 10-year-old boy asked him about space during a breakfast meeting in New Hampshire. "In the old days, it was great," Trump replied. "Right now we have bigger problems. You understand that, we've got to fix our potholes." As he took office, the former reality TV star prioritized the state of the roads, but at some point he was steered toward a supportive stance for the space industry, and commercial space in particular.

The Trump administration has since re-chartered the National Space Council, which had been disbanded in 1993 during George W. Bush's presidency, and made Vice President Pence the chair.[13] This was a great victory for the space community. Barack Obama had made campaign promises to reform the council but it didn't materialize during his two terms in office. The National Space Council is backed up by the National Space Council Users' Advisory Group, which consults with the main council on policy matters. The advisory group is stacked full of famous names from American space history such as Buzz Aldrin, politicians such as the governor of Alabama, and a host of major players from commercial space: CEOs of Boeing, Lockheed Martin, the United Launch Alliance; Gwynne Shotwell, the president and COO of SpaceX; and Bob Smith, the CEO of Blue Origin. There are also some lobbyists in the advisory group, including Stallmer.

After the Space Council was reformed, three directives were put forth. One was to get the United States back to the Moon, sending humans and equipment to the lunar surface, and another concerned space traffic management and which governmental department would be in charge of sorting out how spacecraft and satellites operate in the lower Earth orbit without crashing into each other. The third spelled out regulatory reform. "That does not sound sexy at all, but it was incredibly necessary and needed," said Stallmer. "This had to do most specifically with launch and reentry because the rules and regulations that are on the books now don't even cover the type of launch vehicles that the commercial industry is launching. You didn't have the sophistication of the air launch vehicle, the reusable fly-back boosters, you know, the suborbital vehicles that are flying. None of those are covered in the old regulations [like the Outer Space Treaty]. This gave us a crack at updating the regulations, but at the same time streamlining these regulations that were outdated, unnecessary, onerous to the operator."

Stallmer gives the example of SpaceX launching a Falcon 9 rocket from the Kennedy Space Center. Right now the company has to reapply for permission every time it launches a rocket from the facility in Florida. What Stallmer wants is a "cut, copy, and paste" application where SpaceX would just have to change the payload, the time, and the trajectory of the launch. This would cut some of the red tape that private companies hate so much and allow for the likes of SpaceX to handle their ever-increasing number of flights with greater administrative ease. When the directives were announced, Stallmer and the CSF almost couldn't believe their luck. "This was something that we were just working on in an ad hoc manner amongst the association, amongst the CSF members saying if we get a bite at the apple, these are the things that we would like to see change," he said. "And lo and behold along comes the vice president, and they're willing to let us take a crack at this and update these rules and regulations. So that was a huge deal."

But space companies, despite the apparent support of the vice president, don't get everything they ask for, which means Stallmer and others like him spend a lot of time trying to persuade members of Congress that there's more than one advantage to a commercial space program. "The challenge that I have and the challenge that anyone that lobbies Congress has is that the biggest thing for a politician is a four-letter word: *jobs*," said Stallmer. "They want to know how many jobs will come to their district. Historically, the more money you throw at a program, the better that program is from a political sense because the more jobs that it'll create. But jobs isn't the driving factor for innovation and ingenuity all the time. I'm seeing a 40-person company is building a rocket and I won't name any large govern-ment programs, but there's one out there that has about 80,000 people trying to build a rocket at the tune of $4 billion a year. And I don't see that rocket flying anytime soon, dare I say. And I try to choose my words carefully because these people want to kill me when I speak out against them."

Building rockets for the sake of providing jobs enraged the likes of Bill Stone, who was dealt the same rebuke all those years ago when he tried to sell NASA on the idea of fuel depots in orbit. But the commercial companies now have the momentum, the power, and the influence to make sure they are in the ascen-dency regardless of their potential for job creation. Stallmer was talking about the Space Launch System (SLS) program when he mentioned $4-billion-a-year projects. The SLS has become something of a running joke in the space industry, offering proof to the commercial side of the sector that they are the future of space exploration.

Money and political leverage have made corporations in all industries extremely powerful all over the world, and in a coun-try like the United States, where lobbying is so prevalent, the space companies would be crazy not to spend some of their cash on a little political influence. But it's another area where the new space age opens itself up to the worst aspects of terrestrial indus-try. So far it's mainly the launch companies who have the cash and currency in D.C., but new factions of commercial space are in the process of being formed or revitalized. For one, the space tourism dream has been brought back by companies charging less and offering shorter sojourns out of the Earth's atmosphere, and they are tantalizingly close to taking a long waiting list of millionaires up into space.

Selfie Sticks in Space

DENNIS TITO'S WEEKLONG TRIP INTO SPACE IN APRIL 2001 WAS the first example of space tourism, and it opened the door for six more people to travel to the International Space Station (ISS). Space Adventures conducted most of their training in Russia and launched their clients into space using Russian Soyuz rockets. South African entrepreneur Mark Shuttleworth became the second person to take a vacation in space in 2002, and the same year pop star and 'N Sync singer Lance Bass was scheduled to become the third. Although Bass completed the training, the TV producers behind the trip, who wanted to produce a documentary series from it, failed to come up with the $20 million, and the singer was denied a stay at the ISS.[1] The last citizen to fly into space was Canadian Guy Laliberté, and the millionaires and billionaires of the world have had a frustrating few years with no access to space. But there's a new wave of space tourism companies on the horizon, and a huge number of people already on their waiting lists. But will this new form of space tourism make it a realistic endeavor for anyone but the one percent?

Each of the seven people to go into space with Space Adventures paid somewhere between $20 and $40 million for the privilege, a price that excluded all but the most wealthy from a holiday off-planet. But over the years, other companies have worked to put tourists into space for a much smaller cost. The company expected to put the first paying customers in space using a commercial spacecraft is Virgin Galactic. The company had been founded on the momentum of the Ansari XPrize in 2004, and had developed the second version of its spacecraft. But today, the company has only just reached space. The journey to this point has been fraught with setbacks and tragedy.

*   *   *

IN JULY 2007, A FATAL EXPLOSION OCCURRED AT SCALED COMPOS-ites, the company building the rocket engine that would fire the Virgin Galactic SpaceShipTwo into space. The accident, which happened during a routine test and killed three people, delayed development as the company investigated the cause.[2] Despite this setback, Scaled Composites delivered the new White Knight mother ship—which took the main craft high enough to launch—in 2009. The aircraft was twice the size of its predecessor and was unveiled to the world and the media with much fanfare. The prospect of flying civilians into space was taking longer than expected to come to fruition, but enthusiasm had not waned.

In 2009, Adam Cohen read about Virgin Galactic in the news.[3] You only need to speak to Cohen for 10 minutes to realize he's both extremely enthusiastic about space and prone to acts of severe spontaneity. This explains how, upon learning about Virgin Galactic, he immediately wrote a check for $200,000 to book passage on the company's next available flight. "It was really unjustifiable for me at the time," Cohen told me from Florida via phone. "I wasn't too far into having started a business, so it was not a reasonable thing to do. But space flight is not being powered by people doing reasonable things. It's being powered by people having a lot of passion and optimism who wind up putting what they can into it." When Cohen signed up for a seat in the spacecraft, he was running his own research service for hedge funds and mutual funds, an incongruously boring career choice for an eccentric man. He says the most logical next step in his career would have been to set up his own hedge fund, a reliably lucrative choice. However, fate and spontaneity took him down a different path.

Cohen hasn't complained that he's still waiting to take a trip he paid for in 2009. In July 2008, Branson said the first flight would take place within 18 months. The next year, the company said the maiden flight would be within two years. In December 2009, Branson told a group of customers that flights would begin in 2011. The promises kept coming, and so did the delays. Christmas Day 2013 was mentioned as one possible date, but SpaceShipTwo, now called VSS Enterprise, was not ready.

But none of these delays hurt the company like its next truly tragic setback in October 2014. It was the fourth powered flight of VSS Enterprise, and the vehicle broke apart as it returned to Earth, killing copilot Michael Alsbury and injuring pilot Peter Siebold. The world was shocked at Alsbury's death, and the inevitable questions were raised regarding the safety of the whole project. The accident was later attributed to a pilot error.[4] Still, the likes of Cohen were not put off. "Let's just suppose that they start taking off next year, one way or another, whether it's launch number 200, 500, or 1,000, eventually whether it's a SpaceX delivery of people to ISS or a Blue Origin, one of these is going to blow up just like airplanes do sometimes," he said. "When my time comes I'm going to make a short video that says I'm so excited to go on this trip, and on the off chance that I'm the unlucky passenger, I say it's worth it and nobody should be discouraged by it."

Perhaps it's this attitude, or the willingness to wait no matter how long it takes for the rockets to finally fire, that has forged a community spirit among Virgin's customers. Attenborough describes his 600- to 700-strong customer base as a "pretty interesting group of people" from all over the world. There's also a great sense of duty in what they're doing. "In a way, obviously, I'm very excited to get to go on that trip, but from an almost charitable, philanthropic sense, this is the most important thing I can do," Cohen said. "I write my checks to charities, but anything I can do for space exploration is the most important thing I can do for my great-grandchildren."

This community has come together online and in the real world with a shared sense of anticipation. "There's a private Facebook page for the group," explained Cohen. "They post pictures on that, and when there's a new task or certification we get warned about that. The Virgin people are really good about letting us know what's going on. I think we've got a really good factual understanding of how things progress. Richard, as an optimistic entrepreneur, makes proclamations about when things are going to be done, but whether it's a guy like Branson or Musk, guys who innovate, they always have a timetable that people can't keep up with. I always feel like we know pretty much what's going on, and when it's time we're off to the races." In turn, the customers have become staunch defenders of the concept. "This is like six, seven hundred people who have formed themselves into an amazing community, and are incredibly emotionally engaged with the project. They are our best ambassadors on good days and bad," Attenborough stated.

Cohen says the would-be passengers serve as "kind of diplomatic missionaries" for the Virgin Galactic experience and space travel in general. Virgin Galactic sponsors events within the community that seek to introduce more children to science and technology, and the company also plans to reward their customers' loyalty with additional adventures. One of these was a trip to the National Aerospace Training and Research Center (NASTAR), a training facility outside Philadelphia, where the ticket holders were treated to three days of astronaut training exercises, such as the centrifuge, where a person is spun around rapidly to simulate high g-forces.

Although the sense of community and weekend events are fun, the group is really waiting for one thing. And toward the end of 2018, they took a giant step closer to their goal. On December 13, at 8:03 a.m. on the East Coast of America, Virgin Galactic tweeted: "SpaceShipTwo, welcome to space." The latest test of the Virgin space tourism program had succeeded. In 45 minutes, the spacecraft had gone from Earth to space and back to Earth again, prompting Branson to declare his aim to be on a test flight himself in 2019.[5] The public—and Cohen—would get the chance soon after.

Virgin Galactic's first flight into space with passengers will be a landmark moment for the commercial space industry, but customers will always want to go further into space for longer periods of time. Luckily for them, Virgin Galactic isn't the only player in the game.

*   *   *

THERE ARE THREE MAIN CATEGORIES OF SPACE TOURISM—ONE is suborbital, which is what Virgin plans to offer. The second is orbital, which, as the name suggests, involves entering the Earth's orbit, similar to what customers of Space Adventures experienced. The third type goes beyond that, to the Moon and back.

In the suborbital category, Virgin has two main competitors. Jeff Bezos's Blue Origin is developing a reusable suborbital launch system specifically designed for space tourism. The company would take a maximum of six people on board its New Shepard rocket for a brief sojourn into space. Blue Origin is yet to start selling tickets for flights, despite having passed several test launches, but reports suggest it will charge between $200,000 and $300,000.[6] The other major competitor comes from Russia. In 2016, Roscosmos, the Russian space agency, granted permission to the private company KosmoKurs to build a reusable rocket to send tourists into space. The company is years behind the likes of Virgin and Blue Origin, but plans to offer a similar package as its American counterparts: a short suborbital flight for around $200,000 to $250,000 per person.[7]

Orbital space holidays have slowed to a halt. With no room

for passengers available on the Soyuz rockets for many years, enthusiasm has flagged, or other factors have got in the way. British classical singer Sarah Brightman was scheduled to go to the ISS with Space Adventures in 2015 and began training. But she pulled out for family reasons. That meant the Japanese businessman Satoshi Takamatsu was given the option to take her place, but he declined, saying the art projects he wanted to showcase while in space were not ready. He is currently expected to take the trip in 2020. One of the drawbacks to this type of space tourism, besides the dizzying price, is the amount of time would-be space travelers need to devote to the flight. The training for these missions takes around six months and extensive medical tests are required. These factors limit the amount of potential travelers in this sector.

The third tier of space tourism grabs the most headlines, and indeed it is the most ambitious and profitable. These trips are for the dreamers with more money than they can spend and a reckless sense of adventure. They go beyond the ISS and into deep space. Two companies so far have offered trips around the Moon. One is Space Adventures, operating in conjunction with Roscosmos. The experience involves two passengers and one professional cosmonaut heading to the ISS for 10 days so the tourists can adapt to space. A second rocket is then launched from Earth with a Lunar Module, and the passengers will travel from the ISS on their original craft to rendezvous in Earth's lower orbit with this module, which is comprised of a propulsion section and a living area. The two spacecraft together will then fire their engines and travel around the Moon, flying just 100 kilometers from the lunar surface. In 2011, Space Adventures announced one of the two seats on this journey had been sold for $150 million and by 2014 had found the second customer, meaning the trip could embark by 2017.[8] However, extensive delays mean the company is now aiming to complete the trip by the end of the decade. Nothing is known of the passengers who have reportedly paid to embark on this lunar mission.

The other company aiming to shoot tourists to the Moon is SpaceX—and Elon Musk announced his first customer with typical bravado. In September 2018, to a packed press conference, Musk announced that Yusaku Maezawa, a Japanese e-commerce billionaire, would be the first to take a SpaceX trip around the Moon.[9] Musk announced the departure date as 2023, and also revealed Maezawa had bought all the seats on the flight and will take six to eight artists with him to enjoy the experience. Maezawa is the founder of Zozotown, an online fashion mall and his net worth is $2.9 billion, according to *Forbes*. He is an extremely passionate art collector and spent $80 million on paintings by Jean-Michel Basquiat and Pablo Picasso in 2016. He believes that taking artists to space can inspire them to do great things. "If Pablo Picasso had been able to see the moon up close, what kind of paintings would he have [made]?" the website for the Moon expedition asks. "If John Lennon could have seen the curvature of the Earth, what kind of songs would he have written?"

SpaceX plans to use its Starship rocket and ship (previously referred to as the Big Fucking Rocket by Musk and expected to eventually take people to Mars) to take Maezawa and his selected artists to space. The rocket is yet to be built, however, making the 2023 target an ambitious one. But the project, if successful, has the potential to diversify space. Only 24 people have ever flown to the Moon and 12 of them have walked on its surface. All of those 24 have been white, American males. Just this one trip alone, with artists chosen from all over the world and with vastly different backgrounds, would change that dramatically. This is one of the major reasons to pursue tourism in space—to democratize the experience, make it more readily available for all, and to expand the thinking of a whole generation of people.

* * *

THE DEMOCRATIZATION OF SPACE IS ONE OF THE MOST ADMI-rable goals of space tourism. Yes, these companies are driven by profit, but if they can provide a steady and more affordable route to space then more and more people will be able to experience the revelation of seeing the Earth from outside its atmosphere. Virgin Galactic has made a concerted effort to diversify the people journeying beyond the skies on their spacecraft. "We were keen to try and get that maximum geographical diversification, particularly in the initial group," says Attenborough.[10] "So in those first months of commercial operations, you are having many people, who are the first astronauts for their country. And so that message is taken to the far-flung corners of the globe very quickly."

But the average household income in the United States is around $59,000 per year, meaning a ticket to space with Virgin Galactic costs over four years' annual salary. Income inequality, raging around the world, means this trip will be only for the one percent until the price drops dramatically. That means the vast majority of the world will miss out on the overview effect, a deeply human experience, and an often-cited way space travel can make the world a better place. The phenomenon is said to be experienced by astronauts when they first look down on the Earth from above, when they appreciate just how small and fragile our planet is, see countries with no borders, and witness the whole world rotating as one.

The phrase was first used in 1987 by the author and space philosopher Frank White. In his book *The Overview Effect: Space Exploration and Human Evolution*, he described it as a "cognitive shift in awareness" linked to "the experience of seeing firsthand the reality that the Earth is in space."[11] Many astronauts have spoken about the sensation, including the first man to orbit the Earth, Yuri Gagarin. "Circling the Earth in my orbital spaceship, I marveled at the beauty of our planet," he remarked following his historic voyage.[12] "People of the world, let us safeguard and enhance this beauty, and not destroy it." This was in 1961 and clearly the world has taken little notice. The gears of industry have brought the planet's health to its knees, while wars and inequality ravage the population. The people of the world have not safeguarded or enhanced the Earth, and have in fact taken several steps down the road toward destroying it. Other astronauts have added their voices to Gagarin's since. "It is all connected, it is all interdependent," NASA astronaut Sandra Magnus told White in an interview for his book. "You look out the window, and in my case, I saw the thinness of the atmosphere, and it really hit home, and I thought, 'Wow, this is a fragile ball of life that we're living on.' It is hard for you to appreciate that until you are outside of it."[13]

Clearly this perspective on the planet could help humanity significantly. If more people could see the lack of borders in the world, for example, then perhaps the current rush of nationalism could be tamed. If a larger percentage of the population saw the fragility of the planet, then maybe we would feel the threat of global warming more acutely. But while prices to fly into space remain exorbitant, only the extremely wealthy will be candidates for this wholesale shift in perspective. Optimists may say this is still a good thing—that the people who can afford it, the CEOs, the celebrities, the billionaires, wield more influence and power than anyone else, and so to persuade them will have a trickle-down effect for the rest of the world's population.

But trickle-down economics is still a dodgy policy, let alone trickle-down wisdom. And the space tourism experience naturally preaches to the already converted: if you're going to spend $150 million to go to the Moon, there's a fair chance you're already sold on the basic principles of science, and as an extension, don't need a trip into the cosmos to persuade you to look after the planet. Still, there is a depressingly anti-science, anti-knowledge wave passing through modern culture. Whether it's Bible Belt states banning the teaching of evolution or the all-too-common climate change denial, the world needs to fall back in love with science. And the best way to do that is still inspiration, and space tourism is a visceral way to reignite that spark. There's a strong chance Branson's Virgin Galactic will take passengers to space in the next two years, and when he does, the world will take notice. School kids may feel that they don't need to be top of their class to go to the stars, and everyone can dream of being an astronaut again.

Every new generation will have to deal with the repercussions of how the planet has been treated more than the last, but unfortunately young people will be the least likely to be able to afford a life-changing trip to space. This is why Virgin Galactic is considering ways to ensure a more diverse set of people are able to take the journey. "I think we are about democratization, ultimately, but we recognize that the hurdle is still pretty high," said Attenborough. "So there's two things we can do about that. One is that we can find other ways of connecting people to our experience. You could do that in a very spectacular way through virtual reality. So there's that element. The other is to find opportunities to allow everybody the chance to go. So we will be doing that one way or the other. I hope on a fairly regular basis."

A small startup in Finland is working on other ways to democratize space. Kalle Vähä-Jaakkola, an athletic, tall Finnish man with blond hair and an infectiously enthusiastic manner of talking about space, grew up in the small municipality of Punkalaidun in the west of Finland.[14] He was one of just 3,000 people living in the area, best known for farming. There, Vähä-Jaakkola developed a fascination with space. He only received two TV channels in his home, yet that was enough for him to watch the classic sci-fi shows like *Star Trek* and become obsessed with all things off-world. Vähä-Jaakkola studied physics at Tampere University of Technology, and then at Aalto University, both in his home country. But he soon found himself on the outside of the space industry, desperately trying to participate in a meaningful way. Then in 2013, the opportunity presented itself. He had met Mazdak Nassir, a filmmaker who moved to Finland with his family in the 1980s, fleeing the Iranian Revolution. The two hit it off and began talking about ideas for businesses. As a reaction to the disappointing lack of inspirational stories in today's reality TV–led media landscape, the two formed a plan to set up a competition to put ordinary people on the ISS.

The resulting company, Space Nation, sought to inspire people to be interested in space—and by default their own planet too—through an app, a game, and a grand contest to go to space. The company has raised $3.4 million from a crowdfunding campaign, and its app encouraged users to live the life of an astronaut here on Earth, with grand prizes for those who performed best. Regular challenges such as quizzes, physical tests, and more were used to develop the skills needed to survive life in space, and those who produced the best results were to be sent to an astronaut training camp. Only one lucky champion each year would then go to space for real, but that one person could have come from any background and would venture into space regardless of their wealth or social standing. Space Nation also hoped to train people through the app to be useful members of the space community here on Earth.

Space Nation wanted to get the first person into space through its program in 2019, but that—as with all these projects—always seemed like a tricky goal. At the end of 2018, Space Nation ran out of money and the company was declared bankrupt.[15] The news was a blow for the average enthusiast aiming for space, but there's still hope for the company. In March 2019, the company's former marketing manager Hjörtur Smárason said he and his wife had begun the process of restarting the company along with the two founders.[16]

The achievability of space tourism has been in an uncertain place for many years now. Spacecraft need to be tested, infrastructure built, and multimillion-dollar checks banked, but there's an inevitability about this type of space commerce that suggests it's a question of when not if. Cohen, for one, believes he will go to the Moon in his lifetime, but thinks Mars may be too far off. However, if someone gave him the chance to go on a one-way trip to the Red Planet with no chance of return, would he take it? "Absolutely, and my kids know that. I'm like: 'Kids, I love you, but if someone sells a one-way trip to Mars, I'm there.'"

Cohen was at the Kennedy Space Center to watch the launch of SpaceX's Falcon Heavy rocket and saw the two boosters land simultaneously. He immediately signed up for a master's degree in Commercial Enterprise in Space at the Florida Institute of Technology, an expensive undertaking in its own right. Space travel and exploration has always had the potential to change the way the world's population thinks, but it's always been seen as a luxury that few would ever experience, and low on the list of the world's priorities. But in a roundabout way, an upturn in space tourism could refocus the right people on the right problems and make a significant impact back here on Earth. Whether this comes to pass or if billionaires hoard the experience to themselves remains to be seen. The extremely wealthy of the world have already hoarded opportunity, riches, and land, so it would be sad to see space occupied just by the so-called elite class in the future. Whoever makes it up to the stars, one aspect of space travel that is agnostic to the public vs. private divide, and unsympathetic to the billions of dollars at play, is safety. And some say our activity in space has already rendered it an overcrowded and unsafe space for future travelers.

A Cluttered Universe

WHEN SPUTNIK 1 WAS LAUNCHED INTO SPACE ON OCTOBER 4, 1957, the small metal sphere traveled around the planet for three weeks before its batteries died, and two months later it fell back into the atmosphere and burned up. Aside from triggering the space race, it had the major consequence of beginning a trend of catapulting large machines into the Earth's orbit. Now, everyone is doing it, and things are starting to look a little cluttered in the lower orbit of the planet. The results could be catastrophic.

There have been about 5,450 successful rocket launches since that day in 1957, according to data from the European Space Agency (ESA).[1] During that time approximately 8,950 satellites have been put into orbit. Of those satellites, around 5,000 are still in space, though only 1,950 of them are operational. While Sputnik 1 did the decent thing and disintegrated reentering the Earth's atmosphere, there are nearly 3,000 satellites that didn't—they still circle the planet despite not working, making them large hunks of space junk.

The ESA also estimates there are 21,000 debris objects tracked by Space Surveillance Network, making a total mass of 8,100 tons—that's a lot of trash circling the planet. That junk has found its way into the Earth's orbit by various means. Satellites are thought by ESA to make up about 24 percent of the catalogued objects orbiting the Earth. They rarely crash into each other, but it has happened. About 18 percent of space debris is the discarded parts of rockets, jettisoned during missions into space. An example of these would be upper stages, sometimes called space tugs, which move spacecraft in, around, and out of orbit. Once they've done their job, the space tugs are jettisoned into space. They join other more random objects floating in space, such as launch adapters and even lens covers.

The first crash occurred in February 2009, when a privately owned American communication satellite collided with a Russian military satellite at 11.7 kilometers per second.[2] This took place 776 kilometers above Siberia, and both of the spacecraft were, unsurprisingly, completely destroyed. As a result of the collision, 2,300 trackable fragments of space debris were created. Some of them have reentered the Earth's atmosphere and burned up, but others are still floating above our heads today.

But that's just the trackable pieces of debris. There are thought to be more than 500,000 pieces of junk that are about the size of a marble.[3] Most of these are created by in-orbit fragmentation events. These fragmentation events, where larger objects are broken up into tiny pieces, don't usually occur from objects smashing into each other. Of the 290 in-orbit fragmentation events that have been recorded since 1961, fewer than 10 were from collisions. The majority are from spacecraft and space tugs exploding in space. This can happen when there's some residual fuel in tanks or fuel lines, or some other kind of energy source, lingering on board once the object has been discarded in orbit. Over time, this hardware is subjected to the harsh conditions of space, and their parts begin to deteriorate, leading to leaks and possibly the mixing of fuel components, which can trigger an explosion.

Estimates suggest that there are more than a million objects far too small to be tracked, and because objects can travel at speeds of up to 17,500 miles per hour, even the tiniest spec of debris hurtling into the side of a spacecraft can cause significant harm. Space shuttle windows have had to be replaced in the past because of damage caused by mere paint flecks flying through space, according to ESA. Between the satellites, the spacecraft, the disused rocket engines, the shrapnel from exploded engines and destroyed satellites, and countless other sources, the Earth's orbit is a hazardous place to launch a multimillion-dollar piece of technology. And the debris keeps piling up. The number of satellite launches in 2017, for example, was the second highest in history, and it's showing no sign of slowing down.

"It's cheaper, easier to do and there's a gold rush," said Moriba Jah, an associate professor at the University of Texas at Austin.[4] Jah is the type of space scientist who gets young children interested in the discipline. Smooth-talking and dreadlocked, when asked if he wanted to chat about space debris he replied within minutes: "I'm your huckleberry."

Jah is one of the few dissenting voices against the tidal wave of companies looking to put satellites in the Earth's orbit. "Data is the new gold these days, and space is the ultimate high ground. With cheaper access to space, everybody knows that there's money to be made—untapped billions and billions of dollars to be made in space, starting with Elon Musk and everybody else. And so everybody is rushing toward that, trying to get as many satellites in orbit as quickly as possible to make as much [and] to take as much of that market as they can. So you saw what happened [with] the gold rush. I mean, if you look at the history of the gold rush, nobody really cared that much about the environment, people just set up these towns in the middle of nowhere and they exploited all the resources and then made their money. So that's the thing that I see is that people want to make all this cash, but at what cost to the environment?"

Jah describes himself a space environmentalist, and he's built an enviable resume in the industry. His background is in astrodynamics, the science that explains the motion of objects in space. He completed his graduate studies at the University of Colorado and got a job at NASA's Jet Propulsion Lab. There, Jah had the responsibility of landing spacecraft on Mars, a role he enjoyed for seven years. He left for the Air Force in 2006, where he first got into an area called space situational awareness, which involved looking at space debris and deciding what was a threat and what wasn't. He joined the world of academia in 2015, where he says he has more freedom to choose what he can research. Jah has used that freedom to build a map of the debris, called Astra Graph. The data is drawn from several different resources, including the Department of Defense, a catalog of information from the Russian company JSC Vimpel, and Leo Labs, a private company in America also attempting to pinpoint debris and satellites. The map shows an approximation of where these objects are, but with one big caveat—none of the source data agrees. The Russian data disagrees with the American numbers, and the private sector is out of line with the governments. "We don't have something like an environmental protection agency for space, and I care about this quite a bit, and I want the space domain to be safe, secure, and sustainable," said Jah. "Those are my goals, and I attempt to achieve those goals with my research making space both transparent and predictable. I don't want anything to hide in space. Nothing hides, that's my motto."

As if untracked space debris and an increasingly congested orbit stream isn't enough, another complication has emerged over the past decade, one driven primarily by the commercial sector. Satellites have gotten smaller, cheaper, and easier to launch, and it's created an industry based around a practice Silicon Valley knows very well—data collection. Putting a satellite, or network of satellites, in space and gathering data is called remote sensing, and it's big business. In 2013, only 18 remote sensing satellites were launched, according to data from Forecast International.[5] In 2017, 177 were launched. That's an astonishing rise, and clear evidence of the gold rush reported by Jah. But the projections for the future suggest we've not yet reached the peak of the industry. Between 2018 and 2032, the same data suggests 3,979 satellites will be launched into space, representing an estimated $39 billion in production value.

One company that has gained a quick lead in this area is Planet, a California-based startup with venture capitalist backing and a large fleet of small satellites already in space. Planet has 200 small satellites orbiting the Earth—enough to take a photo of every single piece of land on Earth every day. While it may initially sound terrifying to imagine a network of cameras taking photos of every inch of the planet's land on a daily basis, the data collected can be utilized to do much good in the world. For example, the satellites have been used to help fight the massive fires that hit California, track illegal oil trades, monitor the size of gulags, and many other impressive feats. The company's satellites, named Doves, are each about the size of a loaf of bread, and have a lifespan of around two to three years. For its part, Planet is one of the companies that are transparent about where its satellites are, and has provided data to Jah to help with the Astra Graph. But they're among the only companies who have supplied that data. Jah says he's gotten no help from the likes of SpaceX and Blue Origin. "As much as I pleaded with them, and I've met with them about this issue, they said that they're concerned with other stuff first. Which I don't really understand. I've given them every opportunity to work with me on this, and both SpaceX and Blue Origin have all but given a 'piss off' kind of attitude."

SpaceX is about to become a major player in this field of small private satellites. The company filed applications with the FCC to launch nearly 12,000 satellites into space in order to beam internet down to the Earth, and received permission.[6] The first stage would see a constellation of 4,425 probes, and the second stage would see 7,518 more catapulted into orbit. According to ESA, even in a business-as-usual scenario, where the number of new satellites being launched remained constant, the risks of collisions between satellites and debris would rise significantly. But this is not business as usual, and ESA estimates that doubling the number of objects in orbit would increase the risk of collisions by as much as four times. In this scenario, collisions would surpass explosions as the primary creator of more junk.

The number of smallsats, the term used for anything under 600 kilograms, have increased massively in just five years. According to data from Bryce Space and Technology, over 1,000 smallsats were launched between 2012 and 2017, and nearly 350 of those were in 2017 alone.[7] The company also found that over 475 smallsats over that five-year period were launched by private companies rather than governments. The proposal of SpaceX, and other companies like Europe's OneWeb, which has received approval to launch 720 satellites, has prompted NASA to look into the potential consequences of such heavy traffic in space. A NASA report, published in September 2018,[8] suggests that 99 percent of these new satellites would need to be taken out of orbit as soon as they ceased working, typically within five years of the mission ending. NASA tested what would happen if only 90 percent of the large constellations were de-orbited on time, and calculated that there would be about 260 collisions as a result over the next 200 years. If 99 percent were taken down in time, that number would be reduced to 34. SpaceX has previously told the FCC it will deorbit its satellites within five to seven years.

Deorbiting usually involves lowering the satellite's altitude until it gets caught in Earth's gravity and burns up in the atmosphere. Today, a whole new business has popped up around exploring other options. There's a possibility satellites could be equipped with de-orbiting facilities onboard, such as a balloon or a sail. Once deployed, these would increase the surface area of the craft, causing it to be peppered by more particles in the upper atmosphere, dragging it down toward Earth faster. But these kits would be expensive to install on a satellite, and would increase the weight and cost of launch. Yet another option is to send a spacecraft into orbit with the purpose of removing existing debris. In 2018, a project run by a consortium of universities and companies began testing some models for this.

RemoveDEBRIS, which is being coordinated by the University of Surrey in the U.K., and helped by the likes of the French aerospace giant Airbus, has already experimented with one technique to remove a satellite from orbit. The RemoveDEBRIS satellite, which is about the size of a refrigerator, was sent into space with the goal of capturing a small CubeSat satellite with a net.[9] The tiny prey of RemoveDEBRIS had been put into orbit especially for this task, and footage released by the team after the experiment shows the net deploying and wrapping itself around the satellite. In this scenario, the net wasn't attached to the satellite that deployed it, but the next step is to have the removal satellite tow the net and its captured hardware out of orbit. Another phase of the mission will test a harpoon device that would attach itself to an inactive satellite and again be dragged away by the RemoveDEBRIS satellite. The aim of the project is to build a system capable of removing the larger satellites from the Earth's orbit. Not only are they easier targets to track, harpoon, or net, they are also the biggest dangers to other spacecraft and even the ISS. While a solution like this is needed to address existing satellites, only regulation and best practices can reduce the risk of collision for those yet to be launched.

*   *   *

TO LAUNCH A SATELLITE INTO SPACE, THERE ARE A FEW REGU-latory bodies you need to go through. "Here in the U.S., if you are a satellite operator or a wannabe satellite operator, there are basically two sets of rules that you have to live by," explained Jack Wengryniuk, vice president of regulatory engineering at Inmarsat, the owner of one of the largest global satellite networks, during a lengthy call in which I asked him to explain many things more than twice. "One, of course, is the FCC rules, the domestic U.S. rules, and the second is an international set of rules that are called the Radio Regulations, which is actually a U.N. treaty that is governed by a specialized agency of the U.N., called the ITU, the International Telecommunication Union."[10]

Wengryniuk has been working to put satellites into orbit and navigating the regulatory obligations that entails for more than 20 years, and deeply understands the benefits and flaws of the current system. He says companies usually begin with the international requirements, as they take longer to complete. A satellite operator must indicate precise details about the spacecraft, from its orbital position to the frequency it will operate at, also known as the frequency band. These bands are limited, and only so many satellites can operate within them. The system for assigning these is far from sophisticated. "The simplest way to describe it is first come, first serve," said Wengryniuk.

Certain bands are popular because of the technical benefits they give, such as a clearer and better signal, but they soon fill up, and it becomes difficult to find a free slot. But companies have found a way around the system, to ensure they can lay claim to their desired frequencies, otherwise known as spectrum. When the ITU accepts a company's application, they have seven years to launch a satellite into space. During this time they have reserved the spectrum they selected in their application. But many of these companies are launching constellations—fleets of satellites sometimes numbering in the tens of thousands. As it is, the rules state that only one satellite must be launched to lock up all those frequencies. "The current rules basically say once you launch a single satellite, you're done," explained Wengryniuk. "And the ITU was like wait a minute, you're filing for 10,000 satellites, it includes a huge amount of spectrum, and this operator runs just this one satellite and can lock up all this spectrum, this is crazy."

Work is underway to try to ensure that this spectrum hogging can't continue, and that operators must meet certain milestones after the seven-year period. But satellite positioning options are also limited because of the first come, first served policies.

"If you're first up there, you sort of reserve your place and then you try to keep everybody else away," said Yousaf Butt, an astrophysicist and a consultant on international space policy who formerly worked as a foreign affairs officer in the Office of Space and Advanced Technology at the U.S. State Department.[11] "The people that are already there are very eager to make rules. They say, 'Hey, you know, you should think about keeping other people let's say 200 kilometers away in altitude from our constellation.' Because they're already up there, but had they not already been up there, perhaps they wouldn't be arguing."

And these are just the questionable practices that exist *within* the bounds of the application process. In April 2017, a company named Swarm Technologies, founded by a former Google employee and operating out of Silicon Valley, filed for permission to launch four tiny satellites into space as part of its ambitions to put a network of satellites in space to build a network for internet-connected objects on Earth.[12] The papers were filed, the mission described in detail, and the measurements of the satellites submitted. The FCC rejected the proposal, saying the satellites were too small to be properly tracked and could be in danger of colliding with other satellites. In January 2018, via a rocket blasting off from India, Swarm launched the satellites into space anyway. This incident marked the first ever example of an unauthorized satellite launch by an American company. The FCC sent an angry email, informing Swarm that the regulatory organization wouldn't be considering any future applications until an investigation was carried out to find out why the satellites were put into orbit without FCC approval. All communication between Swarm and the satellites was shut down, and future launches thrown into doubt. Later in the year, the CEO of Swarm, Sara Spangelo, talked to *The Atlantic* about the experience, apologizing for what she referred to as a mistake. Swarm had arranged to have the satellites loaded onto the rocket believing that FCC permission would be received any minute, and even cited cases where it had been given after the satellites were already launched. It turns out, this is a Silicon Valley thing.

"The Silicon Valley way of doing things is technically pretty iterative. You might not be quite sure what frequencies you will land on . . . or where your ride is going to come from," Jenny Barna, director of launch for Spire Global, a satellite data company with 50 satellites in orbit, told Spectrum in 2018.[13] "We've had satellites on the launch pad before the FCC approves it, and then they approve it at the last second." In Swarm's defense, the company had confirmed with Leo Labs, the private satellite tracker that also lends data to Jah's map, that the satellites were indeed trackable, and therefore expected the FCC to accept. But the space industry is now watching closely to see how the company will be punished for its unauthorized launch. The early signs suggest there will be few repercussions for Swarm, which would set a dangerous precedent for American companies looking to send satellites into space. Swarm's goal is a noble one—it wants to bring connectivity to rural areas and developing countries at the lowest cost possible—but unauthorized launches could be dangerous encouragement for actors with worse intentions.

Part of the problem is the ambiguity of the laws and regulations. "I wouldn't really say it's so hard and fast as legally binding regulations," said Butt. "So . . . part of the issue is that there is some wiggle room. If you do not want to adhere to the voluntary guidelines, you can get around them." According to Butt, there have been calls from academics and some within the industry to make the guidelines a lot more stringent and to include private companies in the conversation. Right now the commercial sector does have a say, but not one that's commensurate with its growing influence. "Going forward it's transforming from a national- or military-type enterprise to more of a commercial enterprise and actually the physical number of active satellites I believe is possibly right now more commercial than governmental. What that means for me is that instead of merely involving the commercial sector in helping the governments draw up rules . . .

I think the governments' role here, should be to facilitate the commercial entities from across the world to get together and hash out for themselves what they think are the best rules and what should be optional—what should be legally enforceable."

Unsurprisingly, private companies would prefer self-regulation, but that doesn't seem possible in an environment as fraught with danger as space. One thing that will always motivate a company to comply with best practices is a potential threat to the bottom line. This is where the insurance industry could play a big role in the future of space debris. "Typically, there'll be some period after launch that a satellite will be insured against damage, let's say 100 days or so," explained Butt. "But those folks are also a crucial player because if the insurance rates start to go up because there's lots of debris now in certain orbital slots, then that's sort of self-correcting, then you wouldn't essentially need so much legally binding regulations. People would see [insurance] getting prohibitive expensive and try to reign in their debris."

As with all potentially disastrous scenarios, the world probably won't take notice until disaster actually strikes. "I think it will get serious after there's a couple of serious accidents, it's just the way humans work," says Butt. "We will continue business as usual until then. Deepwater drilling was fine until the Deepwater Horizon accident happened and everybody was fine with the space shuttle until it didn't work so well. So things will presumably continue as usual until things start to go south."

The theory behind the worst disaster imaginable in the Earth's lower orbit is called Kessler syndrome. Named after Donald J. Kessler, a NASA scientist who proposed the model in 1978, Kessler syndrome describes how the space debris problem can cause catastrophic levels of damage. The idea starts with one collision, for example, two satellites hitting each other. That collision would create thousands more pieces of space debris, which would fly off into different orbits. Those pieces could then cause another collision or explosion, which would result in another few thousand pieces of debris. The result is a self-sustaining whirlwind of debris, making its way around the planet at high speeds and taking out a large number of satellites on its way around, and could even affect the ISS. The 2013 movie *Gravity*, directed by Alfonso Cuarón, dramatizes this process in action and highlights the terrifying potential consequences should it come to pass. But even scarier than a Hollywood movie is the thought that the Kessler syndrome may already have begun. The smashing together of the Russian and American satellites in 2009 caused more than 2,000 pieces of debris to shatter off, and although none of them have caused another major incident yet, they're still out there. Kessler himself said in a 2012 interview with *Space Safety* magazine: "The cascade process can be more accurately thought of as continuous and as already started, where each collision or explosion in orbit slowly results in an increase in the frequency of future collisions."[14]

One particular satellite that could catalyze the continuous flow of space debris—ESA's Envisat. The satellite was launched in March 2002 to observe the Earth, and it successfully provided data on a number of different environmental issues, from ozone depletion to ocean temperatures. But in April 2012, ESA announced it had lost contact with Envisat, and it became the largest inoperable satellite in the lower Earth orbit. Weighing in at eight tons, should it be destroyed in a collision it would generate a huge amount of debris.

"Envisat is a very large target, operating at an altitude where the debris environment is the greatest and likely to increase," Kessler said in the same interview. "In the next 150 years that the satellite will remain in orbit, it will become a significant debris source and could easily become a major debris contributor from a collision with debris as small as 10 kg."

Lower Earth orbit is a crowded lane, full of dead satellites, disused rocket parts, and debris from past collisions and explosions. Yet the space industry is pushing ahead and sending thousands more objects up there. The newer, smaller satellites will limit the amount of damage that can be done, but at some point we need to reappraise the system, the regulations, and the laws that govern this part of space. The rush to make money has created a situation in which an unstoppable storm of debris will already be hurtling around the planet by the time anyone imposes meaningful restrictions.

This topic highlights the flaws of the Outer Space Treaty. The document states that any damage done in space is the responsibility of the country from which the mission launched. But the situation in the lower Earth orbit could be so chaotic there would be no way to lay blame on any one party. If one satellite crashes into another and causes the chain reaction leading to Kessler syndrome, is that company and country responsible for the years of subsequent damage the knock-on effects would cause? When the Outer Space Treaty was written, there were only a handful of objects in space, so it should be no surprise it is ill-equipped to deal with the swarm of satellites buzzing around the Earth today.

The proliferation of new technology being placed in the Earth's orbit isn't the only issue here, another factor is that the technology is coming from new places. Even when they do come from governments, it's not just Russia and America any longer. The world's third-largest superpower in space is now China, and the Chinese space agency has a history of making questionable decisions in the lower Earth orbit.

8

The Newcomers

IN JANUARY 2007, CHINA TESTED ITS CAPABILITY TO SHOOT down a satellite from Earth. In a move widely condemned around the world, the Chinese government launched a kinetic kill vehicle—essentially a weapon catapulted into space to smash into and destroy another object—directly at a weather satellite, hitting it, and breaking it up into tiny pieces.[1] This one incident is said to have increased the amount of trackable space debris by 25 percent, according to the European Space Agency (ESA).[2] It was a flamboyant way to announce itself as a major player in the space game, but hardly ideal. In 2019, though, China managed to do something not even the Americans or Russians had done, despite their considerable head start.

On the morning of January 3, the Chang'e 4 lander, named after the Chinese goddess of the Moon, touched down on the far side of Earth's largest natural satellite at 10:26 a.m. Shanghai time.[3] It was the first-ever soft landing on the side of the Moon not viewable from Earth, and opened up a host of scientific possibilities. It also reminded the world that China's space program is one of the most advanced in the world, and it is making the kind of steady progress that could establish it as the leading nation in space in a few decades' time. Emerging from the shadows, a robust space program from the most populous country in the world could reignite the government-led space race, introduce a number of new private companies into the fold, and even speed up various missions to Mars.

The Chang'e 4 mission was the result of meticulous and patient planning. NASA and Russia have never landed a human or robot on the far side of the Moon, primarily because communication is nearly impossible. When Apollo-era astronauts reached the other side of the Moon during orbits, they'd lose contact with Houston, immersed in the peace and tranquility of space while observing views of the cosmos never seen before by humans. The far side is also sometimes referred to as the dark side of the Moon but that's not accurate. The rotation speed of the Moon is synched with the Earth's—a result of the gravitational forces applied during the 4.5 billion years the Moon has orbited the planet—such that one side of the Moon is always facing away from Earth and can't be seen. That there could exist an unknown half of a celestial body so close to Earth has led to many myths and legends and claims that it is permanently dark. But the opposite half of the Moon does in fact get as much light from the Sun as the half we can see. When the section of the Moon viewable from Earth is dark and not visible in the night sky, the other side is completely lit up.

The difficulty in exploring the far side is that communications between Earth and a spacecraft are blocked by the Moon itself. China remedied this in May 2018, when it launched the Queqiao, or Magpie Bridge, satellite into an orbit around the L2 point, a position around 40,000 miles from the Moon that allows the satellite to stay visible to both the Earth and the far side of the Moon. The Magpie Bridge, which is the first satellite in this position, acts as a relay, allowing the Chinese space agency to send signals to the Chang'e rover and its lander. But the nature of the satellite link means there is a delay between mission control and the spacecraft. That delay made landing with remote control impossible, and forced the space agency to develop complex algorithms and automation technology so the lander could navigate the touch down itself.

Xinhua, the Chinese state-owned press agency, described the landing: "After the Beijing Aerospace Control Center sent an order at 10:15 a.m., the Chang'e-4 probe, launched on Dec. 8, 2018, began to descend from 15 km above the moon with a variable thrust engine being ignited, said the China National Space Administration. The Chang'e-4's relative velocity to the moon was lowered from 1.7 km per second to close to zero, and the probe's attitude was adjusted at about 6 to 8 km above the lunar surface. At 100 meters up, the probe hovered to identify obstacles and measured the slopes on the surface. After avoiding the obstacles, it selected a relatively flat area and descended vertically and slowly."[4]

The probe landed in the Von Kármán crater in the South Pole–Aitken basin. The basin, thought to be created by an ancient collision, is the largest and oldest impact basin on the Moon, according to NASA. The diameter of the basin is 2,500 kilometers, nearly a quarter of the diameter of the entire Moon, which is just under 11,000 kilometers. The gigantic hole is also more than 8 kilometers deep, and it's believed the collision that caused the basin to form exposed the lunar crust and mantle materials.

The Chang'e 4 then began its scientific work, some of which were joint projects with international partners. The module carried a lunar biosphere to the surface of the Moon, an experiment that contained plant seeds and silkworm eggs to study how organisms grow in the low-gravity environment of the Moon. It also brought with it a low-frequency radio spectrometer that enabled studies of the sun's atmosphere and can be used in conjunction with the Magpie Bridge relay satellite to form a radio telescope. The radio telescope is a joint effort between Chinese and Dutch researchers, while German scientists also installed a particle detector on the lander.

After the successful landing of the Chang'e 4 module, the rover, named Yutu 2, or Jade Rabbit 2, exited the craft and began preparing for its mission objectives, which included deploying an ion detector, which was a joint project with Swedish researchers. The main scientific aim of the mission was to study the ancient craters within the basins, to establish the rates of impacts on the Moon—and Earth—in its early history. Any analysis of the objects hitting the Moon and Earth in their early years could reveal a lot about the origins of life. The rover trundled off into the Von Kármán crater, an area of the Moon with major significance to China's history in space, as it was named for the Hungarian-American physicist Theodore von Kármán, who was the PhD advisor to Qian Xuesen, a man often described as the founder of the rapidly ascending Chinese space program.[5]

* * *

QIAN WAS BORN IN HANGZHOU IN CHINA IN 1911, AND IN 1935, when he was 24 years old and had just graduated from Shanghai Jiaotong University, he moved to the United States to attend MIT, using scholarship money to fund his trip. He studied at Caltech the next year for his doctorate, and it was there he encountered Von Kármán, who described Qian as an "undisputed genius."[6]

When the United States joined World War II, Qian worked with American scientists to study jet propulsion and produce missiles that could counter the German V1 and V2 rockets. He also participated in the Manhattan Project, the program that developed the first atomic bomb. In 1949, as a result of his efforts and discoveries during the war, he was made the first director of the Florence Guggenheim Jet Propulsion Center at his Caltech alma mater. But developments back in China would soon cut short his stay in America. That same year, communists became rulers of the People's Republic of China, and Qian was almost immediately wrapped up in America's Red Scare and accused of being a communist sympathizer. He applied for U.S. citizenship and was denied, and then applied to leave America, only to be detained. In 1955, Qian was allowed to move back to China.

"It was the stupidest thing this country ever did," former Navy secretary Dan Kimball said later. "He was no more a Communist than I was, and we forced him to go." When Qian returned to China, he was greeted as a hero. He told American reporters as he left: "I do not plan to come back. I have no reason to come back. . . . I plan to do my best to help the Chinese people build up the nation to where they can live with dignity and happiness."

Qian was allowed to establish the Institute of Mechanics in Beijing, and worked within the Chinese Academy of Sciences. His help led to the Chinese testing an atomic bomb in 1963 to 1964, initiating the country into the elite club of nuclear armed nations just 15 years after that dubious fraternity was founded. The work he did in the sciences, at a time when most scientists shied away to avoid the political turmoil, is credited with helping the Chinese space program take off many years later.

In 1970, the first of the Long March series of Chinese rockets put the Dong Fang Hong I, the country's first successful satellite, into orbit. Several launches followed into the 1980s, when China found a new way to fund its space ambitions—selling its services to countries and companies that needed a ride into orbit. "China has, in effect, tried to commercialize its space program since 1988 when it put the Long March launchers on the world market," Brian Harvey, a space analyst and author of the book *China in Space: The Great Leap Forward* told me.[7] "It had to get special permission [from] the United States in order to be able to do that. And that was provided by the George Bush Sr. government, which permitted non-Chinese countries to fly their communication satellites, mainly on the Long March 2E, which was the rocket that the Chinese used for that purpose."

China provided rocket launches for Australian, American, and other satellites, generating revenue for its space program into the early 1990s. Then the Long March rocket program encountered obstacles. The Long March 2E had pre-launch problems. Harvey says subsequent analysis revealed that the specification requirements of the international clients caused instabilities and the rockets subsequently blew up. Things got even more tricky for China after the Intelsat 708 satellite launch of 1996. The American telecommunications satellite was due to be launched on the new Long March 3B. But the rocket veered off immediately after launch and crashed onto a nearby village, killing at least six people. According to Harvey, the Americans accused China of taking advantage of the investigations into the crash to steal Western satellite technology.

The repercussions were severe. U.S. Congress reclassified satellites as munitions, subjecting them to International Traffic in Arms Regulations (ITAR). This effectively ended the launching of American-made satellites on Chinese rockets, and also put limitations on the satellite industry in America. In time, satellite manufacturers in other countries, such as the Europe-based Thales Alenia Space, advertised its satellites as "ITAR-free," as none of their components were manufactured in the United States. But the Chinese space program continued to make strides forward regardless.

In 2003, China became the third nation to put a human into space, when *taikonaut*—the name for Chinese astronauts—Yang Liwei entered orbit in a Shenzhou 5 spacecraft, launched by a Long March 2F rocket. Liwei, a former military pilot, spent just over 90 minutes in space, and marked a major breakthrough for China's ambitions.[8] The country received praise and congratulations from all over the world, including from NASA officials and U.S. president George W. Bush. Further manned space flights were successfully undertaken in 2005, when two taikonauts orbited Earth for five days, and in 2008, when China completed its first spacewalk.

Americans have long feared the rise of China. The country's economy has leapt forward at a staggering pace, and its industrial capacity now dwarfes that of the United States. This mistrust and suspicion has already crept into American policies in space. If China's space agency felt excluded from international affairs after the 1999 incident, it was nothing compared to the ruling passed in 2011 in the United States. The so-called Wolf Amendment banned both NASA and the White House Office of Science and Technology Policy from coordinating in any scientific activity with China. The man behind the bill, Representative Frank Wolf, was a long-time critic of the Chinese government, and at the time chaired a House spending committee that oversaw several science agencies. He said in an interview with *Science Insider*: "We don't want to give them the opportunity to take advantage of our technology, and we have nothing to gain from dealing with them. China is spying against us, and every U.S. government agency has been hit by cyber-attacks. They are stealing technology from every major U.S. company. They have taken technology from NASA, and they have hit the NSF computers. . . . You name the company, and the Chinese are trying to get its secrets."[9]

China was subsequently frozen out of, among other things, the International Space Station (ISS). So the Chinese government channeled its efforts into building their own station. In 2011, Tiangong 1, which translates to "Heavenly Place 1," a small space station, was launched into orbit with nobody aboard. The mission of Tiangong 1 was to help China get a grip on the technologies required to build a space station comparable in size to the ISS. The Shenzhou 8 spacecraft linked up with the station in November that year, and completed China's first-ever orbital docking. In June 2012, the first taikonauts stepped aboard the space station while it was orbiting, before three more visited in June 2013, with both groups staying for two weeks.

Tiangong 1 was only designed to last two years, and by the time the last taikonauts departed in 2013, most of the work in the space lab had been completed. The station met with an ungraceful end, when the Chinese space agency lost contact with it in March 2016. The craft eventually plummeted to Earth and mostly broke up in the atmosphere over the South Pacific Ocean in April 2018. But not before the next stage of the process had begun. The Tiangong 2 had launched in 2016 and was boarded by two crew members that same year. The 8.6-ton spacecraft is another step on the way to China's ultimate goal, a fully functional space station the same size and scope as the ISS. Meanwhile, the Chinese space program continued to serve other countries, but with a new commercial model. "They went to countries in Asia and in South America, and in many of those cases, they not only flew satellites for those countries, . . . they built those satellites and funded those satellites. The kinds of countries I'm talking about are, for example, Bolivia, Venezuela, Pakistan, Laos," said Harvey. "In effect, China was able to get back into the launcher business by selling satellites, launching them, providing a financial package, and indeed a training package for those countries."

China's space program has achieved remarkable progress despite its frictions with the United States and an outright ban on working with NASA. Certainly there were yearlong waits between manned space missions, and most of the goals of the Chinese have been to achieve some of the accomplishments long ticked off by the likes of NASA. But the landing on the far side of the Moon was different. It signaled the moment when the Chinese achieved something America and the rest of the world had not. This could stimulate the competitive side of the U.S. space program, thereby spurring commercial space as well. But the Chinese don't consider themselves to be involved in any kind of race, according to Harvey. "We still tend to look at the space race through the prism and the lens of the 1960s, in which you have two superpowers racing for the Moon, and they were launching rockets and satellites at a mad rate, and every few months there'd be a launch from either Russia or the United States," he said. "That's not how it works now. The pace of things is much more elongated. The Chinese approach has been very much to build up its capacities across a very, very wide range of fronts very slowly, very deliberately with quite a fixed budget that's probably not micromanaged by politicians, unlike the U.S. Congress one.

They've been gradually ticking off all the things that you need to do to build up a really modern space program."

Some in the United States have already issued warnings that the country may not be the major power in space for much longer. Mark Kelly, a retired NASA astronaut, tweeted after the Chang'e 4 landing that the mission is "also a reminder that we need to get back to policy over politics" or "the world might leave us behind." And he's talking only about China's national space program, while the fact is that China's commercial space sector is also coming into its own.

* * *

CHINA'S GOVERNMENT ALLOWED PRIVATE COMPANIES TO BUILD and launch satellites and rockets in 2014. Since then, several companies have attempted to become the Chinese equivalent of SpaceX or Blue Origin, although none have been helmed by billionaires like Bezos and Musk. The companies have instead turned to the venture capital firms of China to get the funding they need.

Cofounder of China Money Network Nina Xiang declared 2018 a breakout year for rocket builders.[10] During that year, several of the big rocket builders in China secured multimillion-dollar investment deals. OneSpace Technology, a Beijing-based rocket company targeting the small launcher market, raised 300 million yuan ($44 million), another Beijing rocket company, LandSpace, completed a 200 million yuan ($35 million) investment round, and the satellite developer Spacety secured 150 million yuan ($22 million) investment. By August 2018, seven Chinese private space companies had together raised more than 1.66 billion yuan ($241 million), according to China Money Network data. The biggest investment of all was a 1.2 billion yuan ($180 million) injection of cash into the launch company ExPace.

SpaceX can't work with the Chinese, due to the contracts it has with NASA. But the way China has developed nascent industries in the past suggests the Chinese government would prefer a homegrown version anyway. And just like SpaceX, many of those companies have their origins in internet companies.

The huge number of internet users in China have always been eyed with a great appetite by the online giants of the west, but the Chinese government has mostly denied them access to the market. The likes of Google, Facebook, and Twitter have been banned in the country completely for many years as part of the Great Firewall of China, named after the technology that blocks access to websites. This allows the government to censor the internet in the country, and also enables huge amounts of surveillance by the state. While censorship may be the chief reason for the ban, another happy consequence for the Chinese has been the emergence of a robust and wealthy set of internet companies in China. In fact, as of 2018, China had 9 of the world's 20 largest technology companies, just behind the 11 in the United States.[11]

In 2018, some of these companies took the first steps toward having a serious space product they can sell. OneSpace and iSpace—not to be confused with the Japanese firm of the same name—designed, produced, and launched rockets into suborbital space, which is a long way down from orbital altitudes, but it's a start. LandSpace did launch an orbital rocket in October of that year, but the small satellite it was carrying failed to reach its target orbit. Harvey describes this attempted launch as "important" for the Chinese space sector, as it marked the first attempt by a private company to put something into orbit, and the satellite itself was also made by the commercial sector. The volume of private companies in the space business has mushroomed in recent years. Nearly 100 startups in the space sector were established in the period between 2015 and 2018, according to media reports quoted by the Chinese News Service.[12] In that same article, it's predicted that the launcher firms in China are now at a critical juncture, as they need to prove they can put payloads into orbit or risk losing future investment. But not all of those

100 companies will be completely private, according to Harvey, who explained that private versus state-run is "very difficult to disentangle" in China.

"The Chinese space program is absolutely full of private companies," he explained. "I think we may have the notion that because they came out of a command socialist economy, you're looking at very big companies in China, and there are definitely some very big companies." Harvey is referring to the level of control the Chinese government holds over the economy. For decades, the government decided which goods would be produced and in what quantities, but more recently the economy has shifted to a mix of free market and communist policies. Because of its history, it's often assumed companies in China are large, government-led entities, but Harvey points out that that is no longer the case—especially among space companies.

"There's a whole set of quite small- and medium-sized ones all over the place. And if you go to the places where China exhibits space products, it's full of large- and small- and medium-sized companies, and quite a few of them are state companies large and small and quite a few of them are spin-offs from the state. So it's a difficult and sometimes confusing picture to look at, and the lines between state and private and local government are actually quite blurred in China."

If you head to the right conference in China, you can see the space industry, in all its complex glory, gather from all over the country. "In Wuhan, there has been a very substantial development with the help of local government, of what's intended to be a very large facility for the development of the private space industry," says Harvey. "They have a conference there normally every November or so at which they invite anyone who's interested in commercial space in China, be they launchers, or be they satellites. And it includes the state sector spin-offs there as well. They've made a very big effort in Wuhan, where they have this huge new industrial facility to try to promote commercial space."

"Our overall goal is that, by around 2030, China will be among the major space powers of the world," Wu Yanhua, the deputy chief of the China National Space Administration, said in 2016.[13] China's spacefaring ambitions come in four main areas: commercial space progress, the Moon, a space station, and Mars. Private companies will need to make significant progress in 2019, but the business ecosystem is growing at such a rate that they will no doubt have plenty of customers. China's return to the Moon is already within reach. The Chang'e 4 mission to the far side utilized the backup vehicles built for the Chang'e 3 mission, which put a lander and rover on the lunar surface, and this next mission will return to the more conventional side of the Moon. Chang'e 5, reportedly scheduled for December 2019, will involve bringing lunar samples back to Earth. These would be the first pieces of the Moon brought back to Earth since 1976. Subsequent missions to the Moon entail looking for ice and other resources, but rumors abound that this exploration program will end with a manned mission. "When the Chinese do land on the Moon, it won't be flags and footprints for 24 hours, 72 hours," said Harvey. "They will probably stay for a month the first time. And then you will have a Chinese base there."

The Chinese space program received the full attention of the global space community in 2019. The private sector still has a lot to prove, but it has raised significant amounts of money and has every chance of catching up with the likes of America and Europe. For now, though, China just wants the respect it deserves. "They want equality with the big space powers," said Harvey. "Being frozen out of the International Space Station really hurt them. They felt really badly upset about that. So they went and built their own. And the reason why they're building their own space station is because the Americans are not permitted by Congress to have anything to do with them. So you've got the superpower politics elements to that, but I always say with the tortoise and hare argument, remember the tortoise does win in the end."

Let's imagine for a moment that China were to put taikonauts on the Moon before the United States or any of its private companies return there. This would be a massive jolt to the American feeling of dominance in space and would have a major ripple effect. NASA would most likely look to its commercial sector to boost its efforts to reach even footing with the Chinese, and in that scramble, the private companies would hold a lot of power over the government agency and its lawmakers.

As we saw in the first space race, when national pride is at stake, a lot can happen in a short amount of time, and should China establish a base on the Moon before anyone from the West, we can expect more American politicians banging the nationalist war drum to get boots on the ground as soon as possible. Budgets would expand, government funding would be increased, and regulations might be eased in this pursuit. China's achievements in space still lag way behind NASA's, but the country has two main advantages: First, their relative political stability means space policy can be formed and maintained over decades, which means no major shake-ups, as seen in the United States every time there is a regime change. Second, the country's private space companies, should they establish themselves, can be more easily orchestrated by the government to aim for the same goals. In the future of space, China is the wild card America created by boxing the country out—to such an extent that nobody really saw them coming.

PART III

THE FUTURE

9

Floating Factories

THE INTERNATIONAL SPACE STATION (ISS) IS, ACCORDING TO some estimates, the most expensive single object ever built. The $100 billion spacecraft orbiting around 240 miles from the Earth's surface is a representation of what the world can achieve when it works together for the sake of science and the advancement of the species. But all good things come to an end, and at some point in the next decade or so, the ISS will be decommissioned or handed over to new owners. That could leave scientists around the world without a base from which to conduct vital studies on space, our universe, and our planet. The commercial space sector is ready to jump in and fill that gap, and at least three companies have serious plans to launch space stations into orbit in the coming years. These floating factories will offer research tools, tourism destinations, and manufacturing capabilities. But a fierce race is underway to be the first one up in space, or to take over the ISS. The winner will become a powerful part of the new economy orbiting the Earth, and will likely change the way we view space stations, from government-run science hubs taking on experiments of national interest to profit-driven factories floating around the planet.

In 2005, the ISS was opened up to the commercial sector

through a declaration by the U.S. Congress. The new desig-
nation of "National Laboratory" meant space research from
commercial and academic users could be carried out alongside
government work.[1] This was the first, small step toward a commercial space station. The ISS National Laboratory was formed
to handle all non-NASA research, and in 2011, the Center for
the Advancement of Science in Space (CASIS) was selected
to manage the research facility. CASIS is a nongovernment,
nonprofit organization tasked with promoting a wide range
of research on the ISS. But that may not be so for too much
longer. The Trump administration has put forward a proposal
that the United States should stop paying for the ISS in 2025
and use some of the money it saves to help deploy a commercial
space station as a replacement.[2] One proposal would see the
ISS partially or completely privatized, with some of the $900
million a year in savings earmarked for the transition to a new
commercial space station. The science and equipment in the
ISS could be transferred to its new home, most likely a space
station that is attached to the ISS for some time, via modules
built by private companies that would eventually detach and
operate on their own.

There are four major companies in the running to either take
control of the space station or launch their own. All of them
have different backgrounds and have taken unique approaches
to fundraising. Not all of them will succeed.

* * *

AXIOM SPACE IS LED BY CEO MICHAEL SUFFREDINI, A FORMER
NASA employee who ran the organization's ISS program for
a decade before he retired from government life in 2015. Suffredini was largely responsible for shifting the ISS away from
government-led science toward becoming a haven for private
research. But even with that kind of experience at the helm,
there's still the question of how you fund as massive an undertaking as the launching of a space station and operating it at a

profit. That task fell to Amir Blachman, who was introduced to Suffredini at an ISS Research and Development Conference in 2015.

Blachman, who speaks as knowledgeably as he does straightforwardly, has a history of investing in space startups and spoke to Suffredini for more than a year before joining the company. He studied Axiom in the same way he would an investment. "I was looking for companies that do something large scale, something that's life improving on Earth, something that moves the needle in terms of international cooperation, a company that is led by people who are really top notch in their fields and not people who are newbies to the subject," said Blachman by phone.[3] He adds that he wants to work for people who are "very well-connected in the business area that they operated because it's just one of the most important ingredients to getting a large-scale business off the ground. And lastly, an ingredient that I thought was very important when I was seeking out which types of companies I want to invest in and work with is the issue of timing, and making sure that the company is creating something that the world is thirsty for now and not 10 years ago or 10 years in the future."

Blachman found that Axiom Space ticked all of those boxes, and he established an investment strategy for a company that would need huge amounts of cash. "The company has two overarching phases to his business. The first is what's going on right now, which is sending our customers to the International Space Station. Simultaneously, we're starting construction of the new station," explained Blachman. "And the second phase of the business will be sending people to our modules while they're connected to the International Space Station during assembly and ultimately to the station when it's separated from the ISS on ISS end of life. And at that point, we'll be operating a standalone commercial station."

The first phase of the plan involves offering countries without established space programs the training and know-how to get a crew up to the ISS. Axiom will offer a wide range of services, from astronaut selection and training to advice on how to ensure that a space program would benefit a country's science and technology education systems. This revenue will provide the bulk of the funding for the space station. "We see those milestone payments well in advance and . . . the business is customer funded in that sense," said Blachman. "So we don't need to raise any more capital for the business of the missions to ISS. Really, the only capital we're raising is to partially fund the construction of the station, which to a great degree is funded by customer proceeds as well." He also says the funding to build the space station will be 25 percent from the sale of equity in the company, about 5 percent from debt facilities, and 70 percent will come from customer proceeds. Asked if there will be any government funding for the station, Blachman is adamant there won't be anything directly from NASA, but does say the majority of its customers will be governments, so a large amount of cash will come indirectly from the public sector.

Axiom boasts plenty of experience and expertise, but it is led by ex-government employees and intends to rely on a construction method—building the station on the ground and assembling it in space—that may soon seem dated and overly expensive. Meanwhile, a smaller competitor has taken what is arguably a more innovative approach, although its priorities may take it to the Moon rather than just the space station.

Bigelow Aerospace was set up by the hotel magnate Robert Bigelow, an eccentric yet low-key billionaire who made his fortune in real estate and owns the Budget Suites extended stay apartment chain. Bigelow has been fascinated with space for a long time, and aside from wanting to take his business to the cosmos, has spent significant amounts of money trying to prove the existence of aliens. It was reported by *Forbes* that Bigelow had given an estimated $10 million to fund a now-defunct UFO-hunting group called the National Institute for Discovery

Science.[4] He is also believed to have bought a 480-acre cattle ranch in Utah in 1996, a site that some believe holds an inter-dimensional doorway used by alien shape-shifters. So far, so weird.

But Bigelow has proved he shouldn't be written off because of his eccentricities. He is a self-made billionaire, and in 2000 he made the shrewd decision to license a technology that NASA had developed and discarded. The TransHab, short for Transit Habitat, project aimed to build an inflatable craft that would be cheaper than the ISS and could be used on a journey to Mars. When U.S. Congress cut funding in 2000, the project was shelved out of fear that another type of module would distract from the over-budget and behind schedule ISS.[5] Bigelow swooped in.

By 2006, Bigelow was ready to launch a version of the technology, an inflatable craft called Genesis. Launch prices in the United States at the time were too high, so the company chose a Russian rocket to send its first test craft into orbit. Genesis I inflated to 14 feet long by 8 feet in diameter and successfully orbited 350 miles above the Earth. Just under a year later, Bigelow was ready for his next test launch and sent Genesis II up into space from Kosmotras Space and Missile Complex near the town of Yasny in Russia. This was also a success, and provided Bigelow with the chance to experiment with a few money-making ideas. Through the company's website, space enthusiasts were offered the chance to put objects into the spacecraft for $299 an item. A wide range of objects were taken to space, including the ashes of loved ones, business cards, and photos. Bigelow also attempted to launch a bingo cage into space (the draw would have been beamed back to Earth via a video link), but the automated arm that would select the bingo balls jammed on the first attempt. All this was leading up to BEAM, the Bigelow Expandable Activity Module, which would theoretically be a habitable spacecraft that could be attached to the ISS. In 2016, Bigelow gathered journalists, industry members, and space insiders at his factory in Nevada to show off BEAM, which by that point was ready to be sent to the ISS.

He stepped up to a podium and began to publicly scold those who doubted him. "You laughed at me," he said to the crowd. "When we said we would build an expandable system and place it on the International Space Station in two and a half years, you laughed," he says. "It's been two years and a quarter. And here we are."[6] In April that year, BEAM was launched to the ISS aboard a SpaceX flight, and was unpacked and installed by the British astronaut Tim Peake. After some early teething problems, the module was inflated a year later, and remains a part of the ISS today.[7]

An inflatable space station is obviously lighter and less expensive to launch into space, but the early designs at NASA ensured it would be completely secure. The inflatable shell of the original TransHab was made up of almost two dozen layers and included superstrong materials like Kevlar to protect it from space debris. When BEAM was sent into orbit, the packed up and deflated module took up 1,400 kilograms of the 2,000 kilograms of space on the SpaceX flight. Although that was a large portion of that payload, it was nothing in comparison to a regular space station module. For example, Tranquility, another ISS module, weighs about 19,000 kilograms. The BEAM module reaches 16 cubic meters in size when it's completely inflated, the size of a small bedroom, or a New York City apartment.[8]

The Bigelow module has been so successful that NASA has decided to keep it attached to the ISS for at least three more years, time that will allow the company to make tweaks and figure out how it will work on a grander scale when it sends a standalone space station into orbit. NASA is currently playing host to the technology and allowing it to develop, but once Bigelow strikes out on his own, the agency will once again become a customer, essentially leasing the BEAM for research. The company's next space station modules are called B330s, named because they have 330 cubic meters of space, and Bigelow aims to launch two of them in 2021.[9] These are standalone modules that are capable of housing six people, and they will go up to the ISS, be inflated as a module there, then detach and become their own self-sustaining space station. In 2018, Bigelow Aerospace announced the formation of a new company, Bigelow Space Operations, to handle selling research room on the modules and even selling the modules themselves.[10]

Blachman at Axiom Space believes Bigelow's inflatable space station modules have an important part to play in the lower Earth orbit economy, despite them being something of a competitor. "We actually hope that they'll succeed because, you know, that's something that we would have use for going forward. So our success is also Bigelow's success in that sense. It's just going to take a longer time to have this available because there's so much yet to be developed around the expandable concepts."

So far, commercial space stations have been closely linked to governments around the world. The ISS sits at the heart of the process, and the business plans of both Axiom Space and Bigelow Aerospace rely heavily on government contracts, whether that's selling their services or, in the case of Bigelow, licensing them to then develop and sell. Other companies have built businesses gradually out of public projects, but now firms are entering this part of the space economy with no previous experience and business plans that are far more uncertain.

Frank Bunger is the CEO of the Silicon Valley startup Orion Span. He tells a familiar tale of how he got the idea to take his entrepreneurial skills off-world—he was disappointed by the government's lack of progress in space and felt the need to make a move himself.[11] That move is the Aurora Station, which Orion Span describes as a luxury hotel 200 miles above the Earth. "I think what got me excited in the last 15 years is that we saw the advent of organizations like SpaceX and Blue Origin," Bunger told me. "When they started, people thought they were crazy. What they have proven is that it is possible to be a privately owned and operated rocket launch company. So what we see with that trend of privatization of space is that it started with the logistics part, but what's coming next, what I felt the time is right for is the privatization of the destination. Aurora Station is a destination for tourists, professional astronauts, and researchers."

Bunger has set an ambitious timeline for this plan. The company wants to send its first customers up to the fully functional space station in 2022. Pricing for the 12-day trip will begin at $9.5 million per person and can be secured with an $80,000 deposit. That price includes a three-month astronaut training program. Bunger has no prior experience in the space industry, and his most recent full-time position was with a software automation company, but he's confident he can get the space station into orbit without the extreme cost associated with government projects like the ISS. "Unlike government we have a profit motive here," he said.

Orion Span plans to send the space station module up into orbit fully constructed, so no work will need to be done on it once it's up there. The plan is to launch it in 2021, and according to Bunger, that version won't be the finished product, but will be operational and will enable the company to start monetization. One of the ways the company hopes to keep costs down is through a modular design, meaning the space station can be expanded whenever there's more demand. "We start with one single launch that goes up and starts generating revenue. When it reaches capacity, we launch another of exactly the same type of space station and attach it perpendicularly to the existing one. So the goal is to grow precisely with market demand and not more," Bunger explained.

But here's the problem—the space station doesn't exist yet. Bunger is taking vast amounts of money from people for what some inside the industry believe is a business plan that doesn't add up. "They're not going to be around for long because they are selling something that just doesn't exist," said Blachman. "It costs tens of millions of dollars to buy a seat on a station, and they're purporting to send people out for $9.5 million for 12 days. The math doesn't work. And the technologies that will support a launch cost that is that low, they're still a decade or two away from happening. It's a nice concept. It's fun to see the interest there, but it's not a serious business."

Orion Span has deployed other methods to raise funds besides selling tickets to space. One of them, rather alarmingly, is crowdfunding. The wording of the press release hinted at a company running out of options and turning to space-hungry enthusiasts to make up the difference: "When we introduced Aurora Station earlier this year, we had an incredible, almost overwhelming response. It wasn't a surprise, as there is so much excitement around the prospect of space travel right now. We're thrilled to be opening up this project to the broader public, giving everyone a chance to chip in on the ultimate mission to colonize space and share in this historic moment."[12] The company's crowdfunding will be hosted at SeedInvest, a platform that promises regular people the kind of investment opportunities usually reserved for the venture capital giants and wealthy angel investors. The platform says it vets all of the companies offered as an investment opportunity through industry experts, but they cannot be any better than the VCs themselves, who have a poor enough hit rate to begin with. Worse yet, these would-be investors can put their money into Orion Span, but it won't get them anywhere nearer to space.

Crowdfunding can be a great way to democratize the investment process and allow companies that normally wouldn't attract big money to go straight to the public and ask for the help to get started. But when a company is already relying on the huge deposits of paying customers to get started, a crowdfunding campaign seems like a way to plug the gaps. The premise of crowdfunding is that a mass of people can make something happen that will ultimately benefit them or the world. The average space enthusiast able to afford supporting Orion Span with a couple of hundred dollars is not likely to buy the multimillion-dollar ticket when the program is ready.

The barriers to enter the commercial space station game are still prohibitively high, and although Orion Span may yet prove people wrong, some parts of the off-world economy are available only to the extremely wealthy or the well-connected. Blachman even believes that lesser equipped companies like Orion Span could do more damage than good. "This is an industry where you still have to have a great deal of expertise and the technology needs to be ready and the relationship with the customer base needs to be there. But most importantly people can't come in . . . without experience and just ignore the numbers," he said. "They're doing something that's actually injurious in that they're asking people to put down deposits for something that isn't going to exist. And raising capital now through a crowdfunding campaign, it's a little bit circumspect and that's not the type of thing that we want to see. We want to see bona fide projects going on that really benefit the reputation of the industry and attract investors to it."

The other entrant in this race is Jeffrey Manber, the man who was there when the Russians first tried to commercialize space back in the 1990s and who ran MirCorp before the space station was grounded. He's now CEO of NanoRacks. Armed with his experience from Russia, he returned to the space station industry in 2009. He'd heard that there was unused space on the ISS that he could utilize and put together a proposal for NASA. "I said, 'I don't want your money,' which got their attention. I want the ability to put research hardware on the station. We'll pay for it, and you'll let us market to whom we wish."[13] NASA agreed, and NanoRacks started with a tiny lab. This arrangement was so revolutionary at the time, Manber didn't know what to charge his customers. First he asked for $90,000 for using the lab on the space station for three weeks, and nobody was interested. He lowered the price to $70,000, and there was still no interest. Eventually he offered the same deal for $35,000 and started to get customers. From there, the company expanded, and has now deployed over 200 satellites from the station on behalf of its customers and has an external platform outside the station. Manber's customers include governments, academics, and others who are unable to secure their research projects on the official ISS payloads. The service the company offers allows more people to get onto the space station and brings the costs down significantly by having all the infrastructure in place and ready to go.

"We are the largest investor in the space station," said Manber. "I have over $40 million committed to ISS and so step by step have shown ourselves, NASA, the industry, how you can have a relationship between the governmental space agency and an entrepreneurial company. Why do I have customers? Because the private sector can sometimes offer services more efficiently and at a lower price than the government."

The next step is for NanoRacks to build its own station, and Manber thinks he can do it cheaper and more efficiently than anyone else. NanoRacks wants to take the spent upper stages of rockets, which are floating around in the lower Earth orbit causing nothing but problems, put them together, and build a space station. The company was able to prove to NASA this was possible in 2018, and is now making its case as the forerunner in the space station sector. "At NanoRacks, we like to say we're the world's first commercial space station company with customers," said Manber. "Because no knocks to my friends at Bigelow and Axiom and other places, but we're in the space station business today. I've got customers, I've got assets, and we're growing the relationship. At the same time we're also investing heavily in how to have commercial platforms that are as economically efficient as we can have, and we're looking at repurposing space hardware."

The diversity of each company's plans shows just how novel

this area of space is. Nobody really knows how to make a commercial space station because it's never been done before. The most serious contenders appear to be Axiom Space and NanoRacks, and both companies are well aware of the difficulties they face. Not only do companies like these fear viable competitors, they fear competitors with a business plan that makes no sense. Too many highly publicized failures and the idea of a factory floating in space quickly move from an exciting opportunity to an unnecessary and expensive pipe dream. It's highly unlikely there will be demand for several space stations (at least initially), so no mistakes can be made, or the replacement to one of the most expensive public projects in history could quickly become a giant reminder of the difficulties of building an economy in space.

* * *

CREATING PRODUCTS IN SPACE IS A COMPLEX AND BURDEN-some undertaking, but in some cases it makes sense on a purely economic level. The perfect example is optical fiber, one of the most sought-after materials in the modern world as a component of telephone and internet lines. Historically copper wire was used to transmit information like phone calls, but optical fiber, which is as thin as a human hair, has almost completely supplanted that technology, allowing messages to travel greater distances with better signal quality. Incredibly brittle, the glass (called ZBLAN) that is intertwined to make these delicate fibers is prone to impurities, which lessen the overall quality. The brittleness of the glass doesn't matter if you're working with it in a zero-gravity environment, as the lessened force has a smaller chance of damaging the product as it's being made.[14]

Made In Space, a California-based 3D-printing company with the goal of printing items in space, has already begun testing its ZBLAN fiber optics production at the ISS. Made In Space has had a 3D printer at the ISS since 2014, which has enabled crews to print out spare parts and tools as they are needed. The idea is that rather than having to send individual parts up to the ISS at great expense, the plans for the parts can be sent electronically and the 3D printer can produce them on demand. Emboldened by that success, Made in Space sent a device to produce ZBLAN to the ISS, as well, and is leading the pack of companies attempting to begin manufacturing operations in space.[15] The company's major competitor is FOMS, which stands for Fiber Optics Manufacturing in Space.

FOMS has also secured a deal to send their own fiber-optic-making device to the ISS, and has the manufacturing payload ready to go, with the flight set for the first part of 2019. The company is confident it will turn a profit fast. "Manufacturing in orbit becomes profitable only for volume production of optical fiber. Fiber is priced per meter and shipped to orbit by kilogram," said Dr. Dmitry Starodubov, chief scientist of FOMS.[16] "The simple accounting estimates indicate that the price of the fiber exceeds the orbital delivery costs by at least an order of magnitude. It is essential to ensure the efficiency of fiber fabrication technology and optimally adapt the process for ISS implementation. FOMS has developed and patented a unique fabrication approach that optimally addresses these requirements. Even the first experimental payload has the opportunity to exceed 40 percent profit margin in the initial flights."

Starodubov says that the company aims to manufacture on the ISS for as long as possible, and he believes the station will be sufficient for the next five years of developing the technology. At the end of that time, FOMS would then consider trying to secure a dedicated manufacturing space at the ISS or a separate, stand-alone platform. Starodubov claims that a FOMS payload weighing 50 pounds could produce 50 kilometers of fiber, and that the full utilization of current payload capabilities will lead to a profitable operation, but when this manufacturing process will afford a dedicated mission, in which the company doesn't share payloads with others, or a dedicated facility in the ISS is "a separate discussion."

Everything from metal alloys to human organs could potentially benefit from being manufactured in space. The key is the microgravity environment. In the case of metal alloys, this lack of gravity allows metals to mix together evenly, allowing metals that cannot be formed on Earth to be created in space. A wind turbine created from a metal alloy made in space will be much stronger and lighter, for example, as it will be made from a combination of metals not possible on Earth. The science behind human organs being printed in space is that, in microgravity, intricate structures that are made of extremely thin materials, like the human heart for example, can be fabricated without gravity forcing the structure to collapse in on itself before it's had a chance to set and solidify.

Blachman says Axiom has already begun discussions and signing formal agreements to work with manufacturing companies looking to produce objects in space. But he believes that manufacturing won't become a significant revenue generator for the company until around 2025. "I think ultimately what you'll see happening on the manufacturing side is that we'll have dedicated areas of our space station and entire modules that are dedicated toward larger-phase manufacturing," he said, adding that once the potential in this area is properly explored, there will be a need for "serious infrastructure" in orbit.

Whether generated by tourism, manufacturing, or government research, there are enough dollars floating around the lower Earth orbit to be sure that there will be several new space stations orbiting the planet in the future. But the ISS succeeded on the strength of some of the largest countries in the world coming together to build something mutually beneficial. This new generation of stations, on the other hand, will be in competition with each other, and that could cause issues.

One person who has concerns is Dr. James Vedda, a senior policy analyst in the Center for Space Policy and Strategy at the Aerospace Corporation, which operates the federally funded research and development center for the space industry in the

United States. "Is there going to be some authority that specifies what the habitation conditions should be? Whether that's temperature range or composition of the atmosphere onboard," Vedda asked during an interview.[17] "Is there going to be some specification for what the docking collars should be, how they should be designed so that there are multiple options for rescue? If they get into trouble up there, you can't have half a dozen different private stations up there and then each one has its own proprietary docking collar and they can't help each other if there's an incident."

A private space station could also create unfair competition among the companies using it. If one corporation took over the ISS or built its own station when the ISS is retired and became the only one in existence, any entrepreneurs relying on manufacturing space inside such a facility would be at the mercy of the only game in town. And considering the huge complexity and costs associated with putting a space station in orbit, it's hard to imagine enough of them being active at the same time to create a robust marketplace.

"A single built-out space station, when it gets to the mature level where you get real capacity to manufacture something in it, at that point it becomes a platform," said Barry Lynn, of the Open Markets Institute.[18] "And if you have 10 companies and they want to test manufacturing in space on this platform and they only have space for like three of them at any one point, then you've got a situation in which the masters of the platform can pick and choose and set deals that give them real control. That's the point where you have a real competition policy challenge. So then at that point you have a challenge that really has now fallen away from NASA, away from DOD, to terrestrial competition authorities." And the international nature of a space station layers even more complexity into the situation. Taking the example of the two competing fiber optic manufacturers, if both were looking for the means to manufacture in space, and a private space station, from Axiom Space, for example, was the only such station, Axiom would hold a great deal of power over the two companies and the global fiber optic industry by extension.

Unfortunately, even the space lawyers aren't thinking about these potential monopolies in space, and ask most antitrust experts here on Earth, and they'll tell you it isn't an area of expertise. The problem, like many potential challenges for private space companies, seems so outlandish nobody is even considering it, despite the fact several well-funded companies are haring toward building their own stations in the next decade. We can compare this to the field of DNA editing here on Earth. The whole concept sounded so much like science fiction that when the first significant steps were actually made, the world freaked out collectively. In the coming years, there's every chance these two pursuits of working in space and editing human genes will be deployed together, as each stands to benefit the other.

The Human Tardigrade

TOWARD THE END OF NOVEMBER 2018, CHINESE SCIENTIST HE Jiankui sparked controversy, debate, and outrage when he announced to the Associated Press and the public through a YouTube video that he'd helped create the world's first-ever gene-edited babies.[1] The claims were later verified—and condemned—by the Chinese government. The breakthrough has immense consequences for the world, the species, and also potentially for the exploration and inhabitation of space.

He, who was trained at Stanford University in America and is an associate professor at the Southern University of Science and Technology in Shenzhen, China, said he genetically altered the embryos of seven couples during fertility treatments, with one pregnancy resulting in twin girls being born. The goal of the experiment was to disable the gene called CCR5, in order to make the girls more resistant to potential infection by the virus HIV, which causes AIDS. In editing the genes of an embryo intended for pregnancy (as opposed to merely for research purposes), He had moved from relative obscurity in his field to a pioneer, and his claims both shocked and appalled his peers.

When He made the announcement, his experiment had not been independently verified and has not been published in a scientific journal. He would not reveal the identity of either the parents or the babies involved and shared a limited amount of data in an Excel spreadsheet at the Second International Summit on Human Genome Editing in Hong Kong. The researcher claims the experiment was self-funded and he had the full permission of the parents. In 2019, China confirmed the existence of the babies, although details are still murky.[2] What is not disputed is that He used the genome-editing technology CRISPR.

This revolutionary technology originated in 1987, when Japanese scientists studying E. coli bacteria noticed repeating sequences in the DNA of the organism. This led to other researchers finding similar repeating clusters in the DNA of other bacteria, and naming them clustered regularly interspaced short palindromic repeats, much more easily remembered as CRISPR.[3]

In 2007, scientists discovered that these clusters served as part of the bacteria's immune system. Bacteria produces enzymes to battle viral infections, and when they kill off an invading virus, other enzymes gather the virus's genetic code and cut it up into tiny pieces. Those pieces are then stored in the fragments in CRISPR spaces, in the bacteria's own genome. Those stored pieces of virus DNA become a reference point for other special enzymes, called Cas9, that search out viruses with the same sequences of the stored DNA. When they find a virus with a matching sequence they chop it up and neutralize the threat. Cas9 is effectively the hit man, while the info in the CRISPR spaces act as IDs of the offending viruses.

A further breakthrough came in 2011, when Jennifer Doudna of the University of California, Berkeley, and Emmanuelle Charpentier of Umea University in Sweden began studying how the CRISPR and Cas9 system worked. They discovered they could trick the Cas9 enzyme by feeding it artificial reference points. The scientists revealed in a 2012 paper they could theoretically feed the enzyme the code for anything, not just viruses, and the enzyme would find it and chop it up. That meant the CRISPR and Cas9 system could be used to attack any part of a gene the scientists desired.

From there, the breakthroughs accelerated. A scientist at the Broad Institute in Boston, Feng Zhang, coauthored a paper showing CRISPR could be used to edit the genomes of mouse and human cells. Around the same time, George Church, a professor at Harvard University, authored a paper with his team showing a separate technique that could be used to edit human cells. Researchers have since found CRISPR can add in replacement genes, as well as cut them out. This opens the possibility of scientists being able to cut out a disease-causing gene and replace it with something more useful. The CRISPR field of science shows no signs of slowing down, and progress is consistently being made with the technology. In 2011, there were fewer than 100 published papers on the topic. In 2018, there were more than 17,000.[4]

It's no wonder, considering CRISPR has opened up a world of possibilities to scientists. It could cost as little as $75 and take just a few hours to alter DNA, and so far the technique has worked on every single organism tested. The technology could be used to edit the genes out of species so they no longer exist, something under consideration for malaria-spreading mosquitoes. The genes would be altered so only males of the species can be born, meaning over time they would die out completely. Or they could add a gene that would make them resistant to malaria in the first place, so they wouldn't spread the disease that kills hundreds of thousands of children in Africa every year.[5] Gene editing can also be used to enhance the nutrients in our food, meaning we could consume less of it, creating more for everyone and reducing the toll that agriculture takes on the planet.

And, of course, it could be used on humans to eliminate genetic diseases, treat current ones, and even create the perfect human. It's this last possibility that alarms those keen to point out when science has gone too far or is "playing God."

Religious objections aside, the ethical argument is valid, and the scientific community has acknowledged the potential horrors of altering DNA. For example, changing genes for the better could produce unforeseen consequences, so a human could be resistant to one disease but more prone to another. Partly for this reason, there is an unofficial international moratorium on using CRISPR on embryos intended for pregnancy. He violated this norm, and suffered international condemnation.

By the end of 2018, He had been placed under house arrest by the Chinese government. It's unclear whether this is due to the nature of his work or accusations that the parents of the twins thought he was testing an HIV vaccine rather than editing genes. The country's vice minister of science and technology Xu Nanping told the state-run outlet CCTV that He's work "crossed the line of morality and ethics adhered to by the academic community and was shocking and unacceptable."[6]

Many of the scientists responsible for breakthroughs in CRISPR technology condemned He's work. Jennifer Doudna released a statement saying: "This work reinforces the urgent need to confine the use of gene editing in human embryos to settings where a clear unmet medical need exists, and where no other medical approach is a viable option, as recommended by the National Academy of Sciences."[7] And Feng Zhang called for a stricter, official moratorium on gene editing in babies.

He himself felt the experiment was a viable one, holding the "real-world medical value" of lessening the susceptibility of the babies to the HIV virus. He also expressed pride in his work and said that it was now up to the international scientific community to decide what will happen next. In one of his YouTube videos, He claimed his research would be appreciated by some outside that community, even if the ethical side didn't stand up for some. "Please remember that while there may be vocal critics, there are many silent families who have seen a child suffer from genetic disease and should not have to suffer that pain again," he said. "They may not be the director of an ethics center quoted by the

New York Times, but they are no less authorities on what is right and wrong—because it's their life on the line."

He did have one major defender, however, and it was a man deeply involved in the development of CRISPR technology—Harvard's George Church. In December 2018, Church said to *Science* magazine: "At some point, we have to say we've done hundreds of animal studies and we've done quite a few human embryo studies."[8] He said it may be a while before the possible side effects and issues reveal themselves, but that shouldn't prevent the science moving forward. "We don't wait for radiation to be zero before we do [positron emission tomography] scans or x-rays." Church has been an integral figure in the CRISPR movement, and is also one of a select few scientists who believe the technology could be put to use making humans more suited to the rigors of space exploration.

* * *

HUMANS AREN'T DESIGNED FOR SPACE. WE CAN'T BREATHE without oxygen and have to contend with poisonous radiation, the challenges of living in zero gravity, and the psychological toll of the emptiness of the cosmos, among other things. There's only one known species on the planet that can survive in space with no form of protection at all—the microscopic tardigrade, also known as water bears (though they don't have to live in water). These tiny animals have eight limbs and are found all over the world, from the deserts, to the mountains, to the seas. In 2007, a European team of researchers sent a group of tardigrades on a trip into space, attaching them to the outside of a rocket for 10 days.[9] When they returned to Earth, the scientists found that 68 percent had survived the trip. Of course, nobody is suggesting morphing humans into tardigrade-like creatures, but borrowing some of their most resilient genes could make us hardier space travelers.

Some humans can naturally cope with the harshness of

space better than others, and one company has even offered to sequence people's genomes to find out if they are destined for a life among the stars. Boston company Veritas Genetics offers a full genome sequencing for $999, and one of the reports it returns to customers details one's genetic disposition for enduring space travel. Among those genes valued highly in space are EPAS1, which allows humans to function with less oxygen, a mutation that causes extra-lean muscles to battle atrophy, and genes that make people good problem solvers with low anxiety.[10]

Theoretically, most genetic inefficiencies can be fixed with CRISPR. One of the most pressing issues spacefaring humans have to overcome, whether they go for commercial or scientific reasons, are the effects of zero or low gravity. On a trip to Mars, for example, humans would have to adjust to three gravity fields. On the way to the Red Planet, a journey that takes around six months, they would be completely weightless. When they arrive on the planet's surface, the gravity would be around a third the strength of Earth's gravity. And when they arrived back home, assuming that was the plan, they'd have to readjust to the gravity here that we take for granted every day.

Moving between different gravitational fields can ravage the human body. Balance, hand-eye coordination, and spatial orientation are all affected, according to NASA, and motion sickness is likely. The space agency, which conducts extensive research on gravitational effects through its Human Research Program, has found that without gravity, the human skeleton suffers some dire consequences. Bones lose minerals and their density drops at a rate of 1 percent per month. Here on Earth, the usual rate of bone loss for the elderly is between 1 and 1.5 percent per year.[11]

There's also the problem of muscle atrophy to worry about. Human muscles encounter much less resistance in zero and low gravity situations, which means that although it's easier to lift things, there's also the danger of losing muscle strength and endurance, something that will quickly catch up with a traveler once they return to Earth. Space dwellers could also experience cardiovascular deconditioning as the body is working less hard in general, and the fluids in the body can shift up to the head, which causes problems with the eyes.

The mechanical fix for these corporeal problems would be to create a gravitational field similar to that of Earth. This may be practical only in a small spaceship rather than on a lunar or planetary settlement, but the theory goes that if you could spin a spacecraft or space station, it would create the gravitational forces needed to form an Earth-like environment inside. But the smaller the vehicle or station, the faster it must spin, so to spin it slowly and comfortably, the spacecraft would need to be huge. It would also be completely disorientating should the spacecraft have windows.

A lower maintenance solution is to use a cocktail of drugs and a strict regimen of exercise to maintain bone density and build muscle, but that may involve working out for large portions of the day just to stay at a steady level of fitness. This is one area where Church thinks the human body could be changed to withstand the effects better, but he admits it's one of the trickier ones.

"We're not only not designed for zero-G, we're not even designed for intermediate gravities and even over the course of a year the impact is incredible," he tells me during a fascinating interview.[12] "It can be done by rotating the platform that would be harder to do on a planet or on a moon, but possible. The other possibility would be to study the physiological systems involved and alter them, but it might be quite a few. And some of them might be hard to change in adults. I think probably you can change just about anything, but it's just a matter of when and how much time and money it will take to make that happen."

Changing the physiological systems of a human, however, would make them more suited to the gravitational field of Mars or deep space, and would turn them into a fish out of water once they returned to Earth. "Well, there are three options," says Church. "You can make it so they're either [best suited to] Earth or let's say Mars. You could make it so you can switch back and forth by like flipping a simple chemical switch. You can make it so that you're adapted to both. It's not ruled out that you could be good at both. There are a few options, the problem is we just don't know enough about what genes would be evolved, and there's a lot of subtle physiological engineering that would be required, probably much more so than any of the other things we've talked about so far, but I think it's totally feasible. My guess is those people aren't going to be going back and forth that much, but I could be wrong. But my guess is probably the most desirable intermediate solution would be something that helps you transition both ways. That would be very popular at the beginning when we are going back and forth a lot."

Living in space can be stressful to the mind as well as the body. The psychological states of astronauts have been closely studied, with the ultimate goal of discovering what characteristics make somebody mentally suitable for a mission to space. Dr. Nick Kanas, emeritus professor of psychiatry at the University of California, San Francisco, spent more than 15 years working as a NASA-funded principal investigator, conducting psychological research with astronauts and cosmonauts. He's written about and studied the psychological issues affecting people living and working in space since 1970. Dr. Kanas says besides the obvious lack of mental illnesses, a member of a team going on a mission to space needs to be able to do two basic things: work alone completing tasks for large periods of time and interact with people during social events like meal times. "You can't have an extreme extrovert or an extreme introvert," Dr. Kanas says.[13]

Psychiatric experts employed by space agencies closely screen candidates for space missions for obvious issues that would affect their ability to do a job, such as claustrophobia, poor social skills, or a short temper. Kanas says it's easy to identify the people who shouldn't go on a mission for these reasons, but it's much more difficult to pick someone poised to really excel.

The screening process will necessarily change in commercial space, particularly when people other than astronauts start making trips off the planet. In the past, space tourists have taken part in rigorous training and have been given tasks to do while they have visited the ISS. But the next wave of space tourists will likely have a different experience. "Number one, you have to have training that's short," says Kanas. "Number two, you have to have something for them to do, so you have to create an experience." One of the major psychological challenges in space is boredom, particularly on long missions and activities would combat that feeling. Despite the fact space tourists will be paying millions for the trips into space, they may still have to be screened before they go, depending on the length of the trip. But Kanas believes that in the future, passenger space flight will be more like passenger airlines, with a crew to look after the customers.

If passengers rocketed beyond the lower Earth orbit to Mars, the psychological hurdles multiply. The distance traveled means it would take months to get back to Earth should they need to, and this can alter the mind of someone effectively drifting in space. Kanas says extreme isolation and loneliness, limited social contact, the delay in receiving messages from home, and the Earth-out-of-view phenomenon would be major stressors on a mission to Mars. If that mission to Mars is a one-way trip, then the psychological makeup of the ideal traveler shifts again. "It might have more adventurous, single, nothing-to-lose type of people versus those who can come back," says Kanas. He adds that if the mission had the possibility of a return to Earth it would attract a more mainstream pool of interested people, though he thinks there would be plenty of people signing up for a mission to Mars, round trip or not.

While genetic issues can be eliminated with CRISPR, the psychology of humans is difficult to alter and more difficult to screen for. The commercial sector will have to vet its employees accordingly, and make sure its customers have enough entertainment space to keep them distracted. The chances of feeling zen in space may become more remote, however, if you know you may be getting sick just by being there.

*   *   *

THE MOST IMMEDIATE DANGER TO THOSE LOOKING TO LIVE
and work in space is radiation. And it's in this area where
Church has made the most progress in his research and experi-
mentation. The dangers of radiation in space have only recently
been fully acknowledged. In August 1972, an enormous solar
flare exploded outward from the atmosphere of the sun, accel-
erating a wave of energetic radiation particles. The wave would
have been dangerous to anyone outside the lower Earth orbit at
the time. It was only due to luck the Apollo 16 crew had returned
to Earth just five months previously.[14]

Now radiation is much better understood and monitored.
There are two main types of radiation that can do significant
damage to humans: protons from the sun and cosmic rays.
These particles, and the secondary radiation they generate, can
penetrate spacecraft, settlements, and human skin, doing lasting
damage to the cells or DNA of the body, and sometimes causing
deadly diseases like cancer.

Here on Earth, humans are protected by a gigantic magnetic bubble, known as the magnetosphere. This bubble deflects
the majority of particles away from the planet, and the Earth's
atmosphere absorbs the few that make it through. The magnetosphere extends out to the lower Earth orbit, which means that
astronauts on the ISS or orbiting the Earth are largely protected
as well, though the station still has barriers and is heavily monitored for radiation levels.

The easiest way to protect astronauts is limiting the amount
of time anyone spends in space. NASA also carefully watches for
solar flares and other events that would cause a spike in radiation, and allows space walks only when the levels are manage-
able. But if companies are to build settlements on the Moon
or Mars, they will need significant correctives to the radiation
problem, as scheduling and monitoring won't provide adequate
protection. Several solutions have been put forward.

One of the better blockers of radiation is actually the most abundant resource in the universe: hydrogen. The easiest way to use hydrogen as a shield is with water. On a long mission in a spacecraft, the water supplies that are kept for the crew could be deployed in a strategic fashion to also act as protection against radiation particles. That could be as simple as storing the water supply in the walls of the vehicle, creating a barrier.[15] This could also work on a base on the surface of another planet or moon. NASA agreed in May 2018 to test an Israeli-built anti-radiation suit called AstroRad.[16] The technology is scheduled to be launched into deep space aboard the Orion spacecraft at the end of 2019, in the first run of the much-delayed Space Launch System, requested by President Obama in 2010, now woefully behind schedule. Orion will carry two dummies, one of which will be naked, and the other will be wearing an AstroRad suit, and on its return they will both be tested to see how much radiation damage has occurred.

Other NASA proposals to solve the radiation problem have been much more drastic. In 2017 at the Planetary Science Vision 2050 Workshop, Director Jim Green discussed building a huge magnetic shield that would enhance the atmosphere of Mars and give protection to the planet and anybody on it from radiation.[17] Green acknowledged that the idea sounded a bit "fanciful," but he says the shield would have other benefits beyond radiation blocking. "A greatly enhanced Martian atmosphere, in both pressure and temperature, that would be enough to allow significant surface liquid water would also have a number of benefits for science and human exploration in the 2040s and beyond," Green said at the event. "Much like Earth, an enhanced atmosphere would allow larger landed mass of equipment to the surface, shield against most cosmic and solar particle radiation, extend the ability for oxygen extraction, and provide 'open air' greenhouses to exist for plant production, just to name a few."

But for Church, the answer lies much closer to home. "You can make cells that are very radiation resistant. You can improve the radiation resistance of bacterial cells by 100,000-fold with just four mutations," he said. "First, we need to collect the data on how you would make human cells highly radiation resistant. That should be about as easy as making the bacterial cells that are radiation sensitive, making them extremely radiation resistant. In both cases, it's just growing cells and exposing them to radiation as a selective agent." Experiments at Church's lab to make bacteria more radiation averse didn't turn up any unintended consequences. He says in the natural world radiation resistance is often accompanied by desiccation resistance—which stops things from drying out—and also freezing resistance, so he's exploring the possibility that the same attributes that make an organism resistant to drying out and freezing make them more protected against radiation as well.

There are ways of changing the biological makeup of the body without editing genes, and one of them solves a problem that is rarely discussed. Anyone working in a space station, a settlement, or traveling on a spacecraft for a long period of time will want to avoid carrying any germs with them. If one person get sick in an enclosed space, it could spell disaster for a mission. A long journey to establish a base on Mars, for example, could represent a great opportunity to make humans pathogen-free. There's an ethical reason to explore this—if humans are to visit other planets where there's a chance there is some life, even on the microbe level, then it could be reckless to introduce our own Earth-based pathogens into that environment. Church, who is a member of the Consortium for Space Genetics, has a plan to test and deploy this. He wants to set up an Earth-based colony that mimics the conditions explorers would find on another planet. While there are examples of Mars base replicas in Arizona, Hawaii, and the Antarctic, Church believes these facilities need to be completely sealed off to enable proper experiments with pathogens.

"Even in earthbound models of space colonies, we could test out many of the things that would make humans different from current humans," said Church, who, despite being a biologist by profession, speaks about space with an un-clinical passion. "For example, this could be tested: having a germ-free environment. That's an option in space and it's an option on Earth colonies mimicking that. We've done this with many animals, including mice, goats, chickens, etcetera. We've even done it with humans. These are humans that had the immunodeficiency diseases, the so-called bubble boy. At least two humans have lived for over a decade each in a pathogen-free environment. It's known how to do it. You just need to do it. And I would say that doing it on Earth is a much safer and more cost-effective way than waiting until we're ready to launch a rocket. You need to do it before the launch because once you've introduced microbes onto a new planet, it's hard to take that back."

Making a pathogen-free environment doesn't require CRISPR or any gene editing, but it could still change the human body, according to Church. He describes it as a way to do human biological engineering without doing genetic engineering, but the outcomes may vary. It could cause the body to suffer a lot less inflammation but become more susceptible to allergies, for example.

Pathogen-free or not, humans are going to be traveling into space more than ever in the next few decades, but how much those spacefarers of the future will resemble us is still to be determined.

* * *

THESE TYPES OF ENHANCEMENTS TO THE HUMAN RACE SEEM like science fiction, but as the Chinese scientist He proved, they could be much closer than we imagine. He may have embellished or exaggerated his accomplishments, or he may even be guilty of misleading his patients, but he has moved the conversation about gene editing in humans forward regardless. For Church, it's a matter of how long this will take to implement rather than a question of whether it's feasible or advisable. "If you're creative, you can save a lot of time and money and my guess is that's how it will go. So for example, for sequencing human genomes, the uncreative approach resulted in a $3 billion genome that took 15 years [to make] and was not of medical quality, so it was low quality and high cost and a long time," he said. "But the creative approach, in just a few years brought the price down to a thousand dollars and it was high quality, it only takes two days now and it only took years to develop the method. So I think the same thing will happen in this case."

And how would the general public react? A poll conducted in December 2018 by the Associated Press-NORC Center for Public Affairs Research suggest that about 7 of 10 people in America would favor the use of gene editing in human embryos to reduce the risk of disease, such as cystic fibrosis or Huntington's disease.[18] Around two-thirds of Americans also favored editing the genes of a human embryo to prevent the inheritance of a nonfatal condition such as blindness or to lessen the risk of developing diseases like cancer later in life.

But about 7 in 10 Americans were against using CRISPR technology to enhance characteristics like intelligence or athletic talent, or for physical enhancements like changing eye color or making someone taller.

The case for using CRISPR to make people more suited to go into space lies in the philosophical grey area between these two scenarios. Space travelers would be at significant risk of developing diseases like cancer if they were to spend large amounts of time outside of the Earth's magnetosphere, but they are also making a choice to do that, and are not at any immediate risk for radiation poisoning from the moment they are born. Church told me that adding the incentive of going to space will make the practice of gene editing only more acceptable to the general public, the scientists, and the ethicists. He also thinks the world will one day accept even the cosmetic changes that editing can provide, explaining that doctors already conduct a huge amount of cosmetic enhancements through surgery.

But a CRISPR-evolved future has another potential consequence. If gene editing reaches the point where it is used cosmetically, or even for working in space, inequality could divide an already fractured population further. Unless these gene editing technologies are offered cheaply to everyone, people already born with enormous advantages could come out of the womb with even more. They could even be designed to excel at a specific job. At the end of 2018, Microsoft founder Bill Gates highlighted gene editing as a trend that not enough people are paying attention to, and warned of the role it could play in inequality on Earth in the future.[19] It's another area where debates on sci-fi-like futures need to occur now before they're all too real and the hour is too late.

And the hour is growing even later. Regulations are barely written down, ethics are still being debated, and all the time the potential to make a huge difference to the human race or, more cynically, to make bundles of cash, means rogue actors like He are only going to become more prevalent. In many ways the world of CRISPR and gene editing bears close comparison to the commercial space sector more generally. Both are young, moving fast, and about to run headlong into legal and ethical debates. Nowhere is that more obvious than the private sector's desire to transfer one of Earth's most destructive practices to our nearest celestial body and begin mining the Moon.

11

The Perils and
Profits of Mining the Moon

WHEN TAKESHI HAKAMADA WAS 10 YEARS OLD HE FELL IN LOVE with *Star Wars*. At his parents' home in Tokyo, he would eagerly await the movies appearing on the family TV, videotapes ready to record them to watch again and again. The films, which portray a great battle for control of a distant galaxy, set young Hakamada on a path.[1] Years later, he heads a company that—along with others—harbors an ultimate goal that has the potential to ignite a scramble for resources right here in our own galaxy. He wants to find and exploit potentially the most promising resource on the Moon: water. The idea of exploiting water on the Moon stretches way back, and has been moved forward by various discoveries throughout the years, including the one Bill Stone stumbled upon in the secret mission control headquarters in Maryland in the mid-1990s. If Hakamada were to get his hands on water on the Moon, it would stoke the already fiery debate about ownership, regulation, and laws in space.

As others before Hakamada have realized—including Stone—finding water on the Moon or asteroids would be a monumental feat because in space, water is fuel. By taking the H from H_2O, rockets could be refueled off-planet, making the Moon a gigantic gas station en route to Mars and beyond. That makes finding water on the lunar surface comparable to striking oil, and would likely prompt a familiar rush for commercial exploitation. Great economic towns, settlements, and companies could theoretically emerge in the desert of the lunar surface, drawn by cheap space travel and the promise of great riches. But throughout human history, resources have sparked conflict. Humanity has fought and killed for water, oil, gold, and precious metals wherever it has settled. With a discovery seemingly in sight, now is the time to discuss ways to avoid that happening again in space. But the laws and regulations are not ready, and entrepreneurs like Haka-mada are pressing on at full speed regardless.

Despite Hakamada's early interest in space, he never harbored any desire to become an astronaut. For one thing, he doesn't believe he'd pass the medical tests required, but he also represents a new wave of entrepreneurs whose dreams and desires for space involve business suits rather than space suits. After studying aerospace engineering at Nagoya University in Japan and then Georgia Institute of Technology in the United States, Haka-mada took a turn away from space. "At the time, even though I studied aerospace engineering, I thought if we ever wanted to commercialize the space industry, we needed money and finan-cial traction. So I decided to gain experience in management and joined a management and consultant firm," Hakamada said on a barely functioning conference call from Tokyo.

After working in consulting for a few years, a grand prize—set up to incentivize the likes of Hakamada to pursue their space interests—drew him back to his cosmic goals. The Google Lunar XPrize, the successor to the Ansari XPrize that spawned the technology Virgin Galactic uses today, was announced in September 2007 to much fanfare.[2] The contest promised a $20 million bounty to the first private company to land a robotic spacecraft on the Moon, have it travel 500 meters, and beam back high-definition images from the surface.

Space enthusiasts heralded the competition as another great carrot for the then fledgling commercial space industry, and teams were formed across the globe to pursue the prize. One of the early competitors, White Label Space from the Netherlands, asked Hakamada to join them in 2008, and he cofounded a Japanese branch of the company. By 2013, White Label Space was moved to Japan, and Hakamada was announced as the new team leader. The team was rebranded Hakuto (named after the white rabbit in Japanese mythology that lives on the Moon) and remained in the race. By this time Hakamada had launched iSpace, and the Hakuto team was part of the larger company aiming to find water on the Moon.[3]

The original Google Lunar XPrize deadlines proved to be wildly optimistic, and the initial target date for a landing on the Moon by 2012 passed without any teams getting close. In May 2015, the organizers extended the deadline again to December 2017, but only if at least one of those teams had secured a contract to launch toward the Moon by December 31, 2015. SpaceIL, an Israeli team, struck a deal with SpaceX, meaning the extension was granted. At this point, only five teams remained, including Hakuto, which had been awarded $500,000 for its lander as part of terrestrial milestone prizes. Hakuto agreed to a partnership with fellow finalists TeamIndus to piggyback off their launch with the Indian space agency. But when the Hakuto rover and lander were ready to go, TeamIndus had run out of money, and Hakamada lost his ride to space. On January 23, 2018, it was announced that none of the teams would meet the latest deadline of March 23, and the prize money of $30 million would go unclaimed.[4]

The failure of the Google Lunar XPrize was a blow to the private space industry. The competitors were not giants like SpaceX and Blue Origin, but smaller, pluckier startup teams attempting to make a significant achievement in space. But Hakamada made sure the competition wasn't for nothing. The Hakuto efforts garnered a lot of public attention in Japan, and drew sponsorship from some of the country's largest companies. This goodwill was leveraged in December 2017 to secure a round of investment worth $90.2 million in iSpace. This cash injection put the company in the position to make it to the Moon regardless of the outcome of competitions, and while it still relies on space agencies and private rocket companies to get them there, its destiny is very much its own.

"It was a very important step for us," said Hakamada about the Google Lunar XPrize. "Without that competition, I don't think we'd be here right now. Even though it was closed without a winner, it created a very important role in this industry. Most of them are not successful, but some of them are still surviving like us. They provided a very valuable environment and protected environment where we can take risks."[5]

Now Hakamada has the money and momentum to keep taking risks, and his company's plans have surpassed a rover beaming back images and video from the Moon's surface. He hopes it will become one of the building blocks of a vast off-world economy. iSpace wants to build a "Moon Valley," where Earth and the Moon become one system supported by an economy in space.[6] Its first mission, which is scheduled for 2019 to 2020, will be the first privately led Japanese attempt to put a lander into orbit around the Moon and relay data back to Earth. At some point between 2020 and 2021, the company will actually land on the Moon, deploying the rover that was built as part of the Google challenge to the lunar surface and transmitting data back to Earth.

Seven missions will follow, all of which are expeditions looking for water on the poles of the Moon planned for 2021. From 2021 on, the goal will be to develop the infrastructure needed to discover and develop resources on the Moon, focusing mainly on water. And the company is not chasing an unrealistic target. Images from NASA's Lunar Reconnaissance Orbiter, published at the start of 2018, appeared to show potentially great stores of ice under the Moon's surface. The pictures suggest that there is an underground network of lava tubes that could provide access to ice, and in turn, water. "At this moment, we have evidence of water on the Moon," said Hakamada. "However, we don't know the total amount of water on the Moon. Or what kind of state the water exists in on the Moon. So we have to explore the potential water area and then find out exactly how the water exists on the Moon and how we can extract it."

In 2017, the company made global headlines when it was reported that iSpace wanted to put gigantic billboards on the Moon. That turned out to be a concept lost in translation. Instead of gigantic adverts carved out in the lunar surface, looming ominously over the Earth at night, Hakamada was merely referring to the sponsorship opportunities offered up by the company. Brands will potentially be able to send their products and logos up to the Moon on a rover or lander, and have them photographed with the Earth in the background. This is one of the many ways the company is trying to raise the funding needed to achieve its primary goals.

Hakamada and iSpace have competition in the race to mine resources off-world. Planetary Resources, a startup based out of Washington, wants to mine asteroids. The company estimates there are 2 trillion tons of water available on near-Earth asteroids, and it plans to locate, explore, mine, and ultimately convert the water to fuel to make space travel cheaper.

Planetary Resources compares the current approach to space travel to taking a road trip from New York to Los Angeles with only one tank of gas and packing every single thing needed for the journey.[7] The alternative in that scenario, of course, is to pick up gas, food, and water along the way. A California-based startup, Deep Space Industries, shares a similar goal, as does Moon Express, a company also born out of the Google Lunar XPrize competition. For all of these companies the Moon, asteroids, and space represent an opportunity to get their hands on untapped resources that will be crucial to future space exploration efforts, and extremely profitable for themselves.

Asteroids aside, the competition for the Moon heated up toward the end of December 2018, when NASA announced it would be working with commercial partners to send small robotic landers to the lunar surface. The companies chosen for the CLPS program, which stands for Commercial Lunar Payload Services, will form a pool of commercial firms that will compete for NASA contracts putting payloads on the Moon. NASA chose nine aerospace companies in total, ranging from giants like Lockheed Martin to smaller companies like Firefly Aerospace, Moon Express, and Orbit Beyond.[8] The CLPS program aims to send small landers capable of carrying at least 22 pounds to the Moon, in an attempt to get more data about the lunar surface. According to the agency, the missions may begin as early as 2019, news that is sure to heighten the eagerness of other private companies to reach the Moon as soon as possible.

Once they get there and resources are established, major questions around exploitation will arise.

* * *

IN SPACE, ONE OF THE ONLY MEASURES IN PLACE TO PREVENT a cosmic gold rush in an environment with no rules beyond its own inhuman harshness is a treaty drawn up over 50 years ago, the Outer Space Treaty mentioned in chapter 1. The treaty has remained unchanged since it was signed, and its language has been more or less adequate to keep spacefaring countries from doing battle beyond the Earth's atmosphere. But now the treaty is on a collision course with the new private companies attempting to make money in space and seeking the fewest regulatory barriers possible.

In May 2017, Texas senator Ted Cruz chaired a hearing of the U.S. Senate Commerce Committee's space subcommittee titled "Reopening the American Frontier: Exploring How the Outer Space Treaty Will Impact American Commerce and Settlement in Space."[9] This meeting was called to discuss possible changes to the Outer Space Treaty, should it negatively impact the plans of companies like iSpace, Planetary Resources, and Moon Express.

"As activities in space increase, they will undoubtedly pose new challenges as countries and companies compete for resources throughout the universe," said Senator Cruz in his opening remarks. "It's incumbent on Congress to use this 50-year anniversary of the Outer Space Treaty to properly determine our actual international obligations." Cruz has called for minor changes to the Outer Space Treaty previously, believing the document is incapable of regulating a space industry that has changed significantly. Even the title of the hearing spoke loudly of the intent, evoking the days of the American frontier, when colonizers spread unchecked over a new land, chasing profits and territory. But surprisingly, commercial companies wanted the Outer Space Treaty to remain untouched. In fact, executives from Moon Express and Planetary Resources attended the meeting and presented their case.

Bob Richards, the chief executive of Moon Express, told the committee that his company supported efforts to streamline regulation in space, but that didn't necessarily mean wholesale international change. "While the Outer Space Treaty may appear antiquated, in some ways it is a remarkably visionary document with profound principles that have served the world well for decades," he said. "I believe time and energy is better spent continuing to interpret the Outer Space Treaty in favor of international collaboration without constraining the rights, the benefits, and the freedoms of U.S. commercial enterprise."

These companies embrace the ambiguity of the Outer Space Treaty because it contains just enough room for more lax laws on a national level. The debate over whether international or national law should dictate ownership of the cosmos provokes hearty discussion among a small yet very active part of the legal community: space lawyers. Frans von der Dunk is a professor of space law at the University of Nebraska and a director of the International Institute of Space Law. Like most in his field, he didn't set out to be an expert on the laws that govern space. After studying modern history at Leiden University in his native Netherlands, Von der Dunk decided public international law would help broaden his horizons and job prospects. Public international law at that time (Von der Dunk graduated in 1987) encompassed human rights, laws of the sea, diplomatic immunities, laws of war, and a tiny fraction of the degree involved space law. Despite having done very little work in the area, he took a position at the International Institute of Air and Space Law at Leiden University in 1990, and eight years later completed his PhD. He published his dissertation on the rise of private enterprises in the European space industry, with his thesis suggesting the easiest way to incorporate private enterprises into international space law was through national regulations.

According to Von der Dunk, an existing section of the Outer Space Treaty coupled with national regulation could keep the private sector sufficiently in check. Articles 6 and 7 of the treaty state that the country of origin for any kind of space activity can be held responsible for any transgression.[10] So if a private company launches a mission from the United States and a rocket ends up crashing into a Chinese satellite, the Outer Space Treaty dictates that the United States is responsible for the damages, as the private company was operating out of its territory. "This is what pushes states, once they allow for private actors to become active in space, to license and control and monitor their activities," said Von der Dunk. "They're under obligation to do so under the space treaty, but they also have a very strong interest in doing so because they would be held accountable." But if every country interested in hosting private space activities drafts their own rules, the result will be a disparate set of regulations for companies depending on where they launch from. "Some [state regulations] are very succinct and very simple and provide for a very heavy regulatory regime. Others are more lenient toward the private sector," explained Von der Dunk.

In November 2015, then president Barack Obama signed into law the Space Act, an update to U.S. law that allows U.S. citizens to "engage in the commercial exploration and exploitation of 'space resources' [including . . . water and minerals]."[11] This law stands on shaky ground, depending on how you look at it. According to the United States, a private entity can extract resources without taking ownership of the land they lie in; other countries disagree. The likes of Belgium, Brazil, and Russia believe that mining an asteroid or part of the Moon is akin to claiming sovereignty of it, yet the United States and Luxembourg argue otherwise. They argue the ban on ownership should be viewed in the same way we see oceans. Under the "global commons" agreement, any nation can fish an ocean, but none can colonize or unfairly exploit them.

In April 2018, the United Nations Committee on the Peaceful Uses of Outer Space met in Vienna to discuss the "activities of international intergovernmental and non-governmental organizations relating to space law." During that meeting, several nations expressed vastly differing views on the topic of resource extraction, including strong pushback against the national laws passed by the likes of the United States and Luxembourg. "The view was expressed that, while proclaiming the universal principles of free access to outer space, and freedom and equality in the study and exploration of outer space, the Outer Space Treaty contained no provisions guaranteeing freedom of action for States, thereby calling into question the stated basis of many national laws on space resource exploitation and utilization," the meeting notes read.[12]

The laws of the seas are often compared to space law, but not always favorably. Maritime law is extremely disparate across the world, which has led to a practice known as "flags of convenience." This refers to shipping companies basing their operations out of certain countries purely because of the lax regulations and laws. If one country's regulations state certain types of ships don't need to have a double hull to help prevent spillages, a company trying to save money and not upgrade its ships could choose to register in that country, rather than spend money on safety measures. That shipping firm then flies the flag of that country, hence the flags of convenience phrase. This could be a concern for the space industry in the future, particularly as more and more companies attempt to break into the sector. But there are a few key differences in the law and the practicalities of space travel that would discourage this practice. First, in space travel of any kind, most accidents are associated with the launch. That means that any country that offers a license and overly favorable regulatory conditions to a space company will likely bear the brunt of the damage, should something go wrong.

On the seas, if a tanker crashes into another ship or grounds itself on the rocks of a country and spills oil onto beaches and wildlife, the law states that the company is liable for those damages. In space, the Outer Space Treaty assigns liability to the host country, not the company. This incentivizes more regulation in space than in shipping, and increases the potential risks should a country gamble on loosening regulations. But the economic rewards for a country hosting space activities are tempting. Space travel is an expensive endeavor that employs a lot of people and attracts wealth to a country. So it's no wonder that places such as Luxembourg are passing light regulation laws in order to attract space companies.

And it's working: Hakamada's iSpace is a Luxembourg-registered company. "I think a lot of companies will go to Luxembourg," said Kyle Acierno, the managing director for Europe at iSpace. "That's why Luxembourg is creating policy in the first place, to attract companies to set up an office. What I think those countries will do is ask them to have a little more than a mailbox, like a support company, in order to have the Luxembourg flag. I don't think it's an issue, I think it's a reality. Countries do this all the time. Think about Apple and its offices in Ireland. It's very common for countries to make special policies to attract companies to them."

Acierno is a member of the Hague International Space Resources Governance Working Group, an organization aiming to establish the building blocks for new legislation in space.

The group is composed of experts from the commercial sector, governments, and academia, and its objective is to remove the ambiguity that surrounds international space law. The group is currently finalizing a plan for a new policy, which would side-step the ban on claiming land on other celestial bodies. Acierno thinks this can be accomplished in the name of safety, with the establishment of safety zones in which one company operates to the exclusion of others.

The lunar surface is covered with a fine dust called regolith, which is very similar to sand. When a vehicle lands on the surface, this abrasive dust is kicked up into the air and because there's no atmosphere, it flies, and according to Acierno, can be sent up to 50 kilometers away. If equipment is set up in the vicinity of another landing, it would be sand-blasted and destroyed. "As you know, Article II says space cannot be appropriated," says Acierno. "The way around that is to have a safety zone, where it's not necessarily appropriating the land it's just saying we're working there. Which is fair, we already have this in space; we have it at the International Space Station, for example, which prevents other satellites from entering that space. It makes sense."

Acierno hopes the safety zone law will be in place for iSpace's first landing on the Moon in 2020. But he isn't expecting, or even chasing, an international agreement. "It won't just be one body, we'll bring it to the U.N. and talk about it there, also to national assembly. So Luxembourg and Japan and probably the [United] States at the same time. If those three countries trilaterally agree, I don't really care what countries that have no interest in space say," he explained.

The safety zone proposal undoubtedly has practical value to anyone operating on the surface of the Moon or another planet, but it has the potential to seriously undermine the Outer Space Treaty. The ambiguous wording of the treaty creates a vacuum to be filled by self-interested interpretations. And the reason nothing serious has been done to address this is because the majority of the international community feel practices like mining the Moon are a long way off. But even if they are, any lag at all between the discovery of resources and the establishment of a coherent set of global space extraction laws will allow companies to act dangerously and selfishly.

One treaty has been proposed that would address any potential commercial activities on the Moon, but the major spacefaring countries haven't been eager to sign up. The Agreement Governing the Activities of States on the Moon and Other Celestial Bodies, better known as the Moon Treaty, was opened for signatures in 1979.[13] The Moon Treaty attempts to clarify and close some of the perceived loopholes in the Outer Space Treaty by banning ownership of any extraterrestrial property by any organization or private person, unless they are both international and governmental, such as a global organization like the U.N. At the beginning of 2018, just 18 countries were parties to the treaty, and none of them have a major space program. The treaty remains in limbo, but should a private company land on the Moon and attempt to exploit its resources, it would suddenly become a very pertinent document. Until then, it seems most countries will wait and see what its private companies can achieve and sort it out, based on their own interests, then.

With gas stations on the Moon, mining operations on asteroids, and Donald Trump's space army poised to militarize space, Hakamada's favorite films no longer seem so far-fetched. The international laws that prevent a dangerous land grab are under threat from workarounds and contrary national interpretations. Should a company like iSpace find water on the Moon tomorrow, it will most likely be ready to stake a claim for the surrounding area the same day. The laws that will prevent any conflict need to be ready. "It's the people that are brave enough to make the first entrance that get to make the rules," said iSpace's Acierno.[14] With that mindset, and with almost complete disregard for the Outer Space Treaty, others are already looking beyond the Moon to Mars.

Mars and Back

MARS HAS ALWAYS HELD A SPECIAL PLACE IN OUR COLLECTIVE imagination. The fiery red planet, named after a Roman god of war, has spawned myths, legends, and a whole genre of excellent sci-fi books and films. Since the turn of the millennium, our obsession has led to concrete plans to reach, explore, and ultimately settle on Mars.

NASA plans to put humans on the planet, and has a long-term vision to make it happen. But the private sector doesn't want to wait, and the fans of Mars are ready to jump aboard the next rocket to the fourth planet from the Sun. These enthusiasts see themselves as pioneers making humanity's next logical step—becoming an interplanetary species. But for the companies looking to exploit Mars, they're just customers.

After the September 11 attacks in New York City in 2001, Bill Hargenrader signed up to join the army.[1] He soon found himself on an airplane flying to Iraq with 200 pounds of gear underneath him, a mission to complete, and a nagging doubt in the back of his mind. Hargenrader had always been interested in space travel and exploration, a fascination owed partially to his love of *Star Wars*, but it wasn't until he was in the Middle East on a military assignment that he had a revelation: "I knew that

I wanted to serve my country," said Hargenrader, "but I also knew that war is not the answer. And so I was conflicted as I was heading to war, but when I served, I was looking out the window at night at the stars in the sky and a thought popped in my head and I wrote it down in my journal—if we spent half the amount of money and funding and resources to get to Mars as we do on war, we would be there by now." In 2014, he reread that journal from Iraq and had the same epiphany all over again. He decided he would dedicate the next 20 years of his life to getting humans to Mars.

Hargenrader, who is well-built and tall, with longish hair, has led an interesting life. When he was eight his brother set fire to his family home and he was blamed, and subsequently spent time in a mental institution. He managed to prove his sanity and make it back home only for his brother to set fire to the house again, and this time Hargenrader was sent to a mental health institution for two years. When he was 13 years old, he returned to his family and eventually his brother confessed, but Hargenrader was obviously deeply affected. He decided then to "become the hero in his own story," and if that sounds like motivational speak it's because it is—he now sells self-help books and gives talks about his early life experiences. All of this makes Hargenrader a perfect candidate to lead a Mars-obsessed community. He has more reason than most to look for inspiration from the world, he wants to do something larger than himself, and he's quirky enough to follow through.

Since pledging to himself to make a difference in the quest to reach Mars, Hargenrader has published a popular science-fiction book, set up a media organization dedicated to Mars, and hosts *Red Planet Radio*, a podcast in which he and a fellow host interview experts from NGOs, commercial companies, governmental organizations, and others about how to reach Mars.

Hargenrader's I Love Mars project aims to bring those disparate groups together in one place. He hosted a five-day online summit on all things Mars, science, and sci-fi, which featured some prominent speakers like Andy Weir, the author of *The Martian*; Leroy Chiao, former astronaut and International Space Station commander; and Neil deGrasse Tyson, the celebrity scientist and author. The I Love Mars community is home to a number of people who are, to put it mildly, obsessed with the notion of going to Mars and believe that not enough is being done to make sure this happens. "I've got friends at this point that I know would literally die for just a chance to try to get to Mars, so let's start coming together and build something around this," says Hargenrader. This kind of fanaticism leads to strong communities, vibrant discussions, and a ready-made customer base for any commercial companies that actually make it to the Red Planet. But it also makes people vulnerable to exploitation, and some projects and organizations, whether well-intended or not, have let many people down.

In 2012, a Dutch organization called Mars One proposed a plan to colonize Mars that involved taking a group of humans to the Red Planet and leaving them there.[2] Mars One is made up of two parts: a nonprofit organization attempting to speed up the colonization of Mars, and a for-profit company that owns the broadcast rights of the reality TV show they planned to make about the adventure. Mars One, unsurprisingly for a project asking people to move to Mars to never return, attracted international attention when it was announced, and for a good few years afterward.[3] The organization was founded by Bas Lansdorp and Arno Wielders. The pair had a vision for a private mission to Mars, which would primarily be funded by broadcast rights to the whole process. Lansdorp was quoted in 2013 saying he needed a budget of $6 billion to put the first four-person crew on the planet, and also claimed that much of the technology needed to put humans on Mars was already invented, or nearly ready.[4] That year, Mars One started accepting applications for would-be astronauts and colonizers.

"One day, I was driving in the car with my friend . . . and she said her cousin from Minnesota had just signed up for this really special project that was a colonization project on Mars," recalled Sue Ann Pien, who until 2019 was still in the running to be among the first set of Mars One astronauts. "The moment she said that I pulled over and turned to her and said, 'Tell me more.' And she explained what it was and then she said it was like a one-way project. Something in me just knew I had to apply. I went home and immediately looked it up and started my application process and really took time to plan it out and go and make my video answering all the questions, and I was really lit up, really excited about it."[5]

When I spoke to Pien in 2018, she was still in the third round of the contest, which had stalled significantly since it was first announced. According to one of the early timelines, and the one still displayed on the Mars One website, the first crew should have begun training in 2018. That year there were still 100 people in the running, with no sign of the next round of cuts beginning in the near future. "There's 50 women and 50 men currently and that makes up the 100 of us that are in the third-round stage. The next round will see us divided into teams of four with two men, two women international teams that are going to go on to train if we make it into the fourth round," said Pien.

Pien comes from a family of scientists. Her mother and father both worked in the aerospace business, and she grew up listening to her father talk about black holes and his theories about space. Pien says she always knew somehow she'd end up in space, and Mars One offered her the perfect opportunity. Pien is also an actor, and the chance to gain worldwide exposure while pursuing her dream of living in space can't have been lost on her. Since sending in her application, she has had to complete various tasks to make it to the third round, where she now waits. She was asked to do a physical examination to check she was healthy enough to go on the long journey to Mars, and says a number of people dropped out at that stage. Mars One said that during this stage many people found illnesses they weren't aware of, from serious issues to conditions like color blindness. After the physical, the questions began. Pien was tested on her ability to memorize crucial information about Mars, something she described as like being in school. But she says the real test will come in the next round, when the groups form and they find out who works well with others. But nobody, Pien included, really knows if the next round will ever happen.

When the Mars One mission was announced, it was criticized by many scientists and business people as being completely unrealistic. And the initial deadlines proved to be just that. Over time, the estimates of when the first Mars One colonists will blast off into space have slipped on multiple occasions; the last update put the launch at about nine years after the original timeline.[6] The lack of funding alone was enough to doubt the mission could ever happen, and the company likely got off to a worse start than it let on. Mars One CEO Lansdorp repeatedly claimed that they received more than 200,000 applications to be part of the project, and each of these paid around $40. That would have been an $8 million head start—but there were many reports that the actual number of applicants was much lower. The Mars One chief medical officer Norbert Kraft once told *The Guardian* he had seen 80,000 applicants, and in 2013 NBC counted the number of video applications on the Mars One website and arrived at a figure of 2,782.[7]

The organization was then unable to strike any deals with television broadcasters, internet streamers, or documentary makers. Dutch production company Endemol, which produced the international reality TV hit *Big Brother*, was initially in talks with Mars One about a show, but revealed in February 2015 that no deal had been reached and Endemol was no longer involved in the project.

Mars One responded by pursuing other forms of funding. In December 2013, the project launched a crowdfunding campaign on Indiegogo to raise $400,000 for the first private mission to Mars.[8] The money would be used to fund a lander, which would reach the planet's surface, and a satellite to be deployed in orbit around Mars, according to the now-dormant, but still visible, crowdfunding page. The organizers aimed to reach the goal by January 25, 2014, but extended the deadline to February, at which point they had reached $313,744. Neither the lander nor the satellite was ever built.

Mars One never made its finances public, but the desperation behind some of the fundraising attempts was clear. In 2016, the for-profit side of Mars One underwent a reverse merger with a Swiss firm called InFin Innovative Finance AG. This was supposed to allow the company to be listed on the Frankfurt Stock Exchange, but the plan didn't work, and the German-based stock exchange didn't allow Mars One to trade its shares. Reporting by Rae Paoletta for Inverse also revealed that the company has been in serious debt for many years.[9]

The problems were not just financial. A report released in 2013 by engineers at MIT suggested the technical aspects of Mars One's proposal didn't hold up. According to the analysis, some of the projections for the project were wildly off. For example, Mars One stated it would need to use six launches of SpaceX Falcon Heavy rockets to send up the supplies before the colonists arrive. The MIT team determined they would actually need 15 rockets. Putting aside the fact that Mars One doesn't have an agreement with SpaceX or any other launch provider, that would make the cost of that part of the trip $4.5 billion, a considerable chunk of the total budget. "We're not saying, black and white, Mars One is infeasible," said Olivier de Weck, an MIT professor of aeronautics, astronautics, and engineering systems, in MIT News at the time.[10] "But we do think it's not really feasible under the assumptions they've made. We're pointing to technologies that could be helpful to invest in with high priority, to move them along the feasibility path."

Instead of taking the ideas of the MIT engineers under advisement, the leadership team of Mars One decided to debate them instead. Publicly. Lansdorp, one of the debaters, offered no insight or argument regarding the technical side of the mission. Instead, he tried to sell the dream of going to Mars at an event where people wanted to hear details.[11] As a result of this, and SpaceX's denial the company was in negotiations with Mars One over the mission—and many other reasons—contestants began to pull out.

The 100 third-round contestants were announced in February 2015, but a month later one of the 100 spoke out about the mission's bleak prospects. Joseph Roche, an astrophysicist at Trinity College Dublin, wrote opinion pieces for *The Guardian* and also Medium, raising major doubts over the selection process, which he said was not comparable to any actual astronaut training, and revealing that despite having made it to the final 100 people in the contest, he'd never met anyone from Mars One in person.[12] Roche said he was originally told he'd take part in regional interviews, and would be tested for several days, but that was downgraded suddenly to a 10-minute call over Skype with Mars One representatives. Roche also called into question the ranking of candidates, saying the process wasn't transparent enough and claiming those put in the top 10 list of candidates had merely donated more to the cause than others. Other candidates, like Pien, stayed with the program despite the reports, even if she is realistic about the chances of the project succeeding.

"It's very important to have divergent discourses on something that is this big and interesting, and you know, that's the nature of the scientific method, right? So I think opening up the discussion of all the reasons why it's not going to work is going to be important because that's how we're going to find a solution," Pien said.

In 2019, Mars One announced that the for-profit side of the organization had gone bankrupt.[13] Although the founder Lansdorp refuses to acknowledge the mission is dead, and retains hope the organization will bounce back, it remains to be seen how Mars One will fund its endeavor without revenue from its for-profit business.

From the start, Mars One gathered a number of advisors from aerospace and other scientific specializations. They gave the project legitimacy, and presumably were offering some form of advice behind the scenes. One such advisor was Dr. Robert Zubrin, a man well-known in the communities supporting and encouraging a manned mission to Mars. Zubrin is a former staff engineer at Lockheed Martin Astronautics, and is now the president of his own company, Pioneer Astronautics. He wrote *The Case for Mars: The Plan to Settle the Red Planet*, which is considered something like a bible for the Mars colonization fan base. He has degrees in aeronautics and astronautics and a doctorate in nuclear engineering from the University of Washington. On the Mars One website, his quote backing the project reads: "Mars is the new world. Its settlement presents the challenge that will determine whether we remain confined to Earth, or can become a multi-planet space-faring species, with a future made unbounded by our courage and creativity. Mars One has accepted that challenge. It is a daunting one, and the odds may well be against them. But if no one tries, no one will succeed. I'm proud to do what I can to help."[14]

It's hardly the most comprehensive of endorsements, but it's one Zubrin has since reversed. "Mars One was always very weak, and frankly I question whether it was ever a serious attempt to launch a Mars mission. I think it was an attempt to launch a reality TV show," he told me.[15] "I don't think Mars One was a correct approach, there is something to be said about a Mars settlement organization, a global organization of people, willing to fund the settlement of Mars. If you look at some of the bolder episodes of colonization, say the pilgrims going to Massachusetts, the Mormons going to Utah, and the Jews going to Palestine in the twentieth century, they were all supported by large organizations that helped fund the actual immigrants," Zubrin said. He suggests such an organization could be set up by the supporters of a Mars colonization plan, arguing that there are around 1 billion people on Earth with some sort of disposable income, of which around 10 percent believe it's important humans expand into space in the future. From those estimations, he concludes that if they were to pay $100 a year each, that would be $10 billion a year, which would be enough to fund a Mars settlement. "Now Bas Lansdorp never attempted anything like that, he was focused on his reality TV show. So Mars does not need a reality TV show, Mars needs a Theodor Herzl," Zubrin adds, referring to the man known as the father of the state of Israel.

He proposes this fundraising model with complete confidence, but persuading 1 million people to spend $100 a year on a mission to Mars would be an incredible feat. Ultimately, it's the same kind of wishful thinking that fueled the Mars One project.

Zubrin says he's not setting up his own Mars settlement organization anytime soon, but he does lead the nearest thing to it—the Mars Society. The nonprofit organization was founded in 1998 by Zubrin and others. According to Zubrin, the organization fulfills the three basic needs in the quest to put humans on Mars. The first is political work to try and defend the various plans to go to Mars and encourage others. The second is the projects the society undertakes itself, like its research stations in Utah (called the Mars Desert Research Station) and the Arctic. The final aspect of the society's mission—which Zubrin says is the most powerful—is general outreach and spreading the vision of going to Mars.

One part of that outreach involves using social media to spread the message that going to Mars is an important task humans must undertake. The Mars Society's Facebook group has over 24,000 members, despite being a closed community where users must be approved before gaining access. The society's page has more than 67,000 likes. The group also boasts over 18,000 followers on Twitter and has chapters all over the world. But Zubrin sounds most proud when describing how the Mars Society pushed Elon Musk toward a Mars mission all those years ago back at the start of the millennium.

* * *

THE SPACECRAFT THAT MUSK WANTS TO BUILD TO SEND PEOPLE to Mars has had many names over the years, but has always been the ultimate objective for the billionaire. Musk has referred to the vehicle on several occasions in the past, although it was only officially announced in 2017.[16] In June 2013, he spoke of a Mars Colonial Transporter, and in 2016 he discussed the Interplanetary Transport System launch vehicle, which required gigantic rockets to move people to other planets and moons. Onstage in 2017, he revealed the BFR, which could either stand for Big Falcon Rocket or Big Fucking Rocket, but either way he got the point across. This was a huge rocket capable of traveling between planets in the solar system. But it was smaller than that proposed in the Interplanetary Transport System, and for good reason. The key to making Musk's Mars plans fundable was to make the BFR as useful for taking people to Mars as it was for putting satellites in the lower Earth orbit, taking astronauts to the ISS, and landing on the Moon. This tweak would allow the BFR to cannibalize all the other rocket programs SpaceX has and use up the leftover resources. Musk subsequently changed the name of the BFR again, naming the second-stage rocket and ship Starship, and the first stage Super Heavy.

Whatever the name, SpaceX is building a vehicle that will have multiple commercial uses near the Earth and the ability to take people to distant planets as well. Musk has set some ambitious goals for reaching the Red Planet. He wants to land the first Starships on the surface of Mars in 2022 with two missions dedicated to carrying cargo. Then, in 2024, SpaceX will put the first people on Mars, according to the timeline, and by 2028 finish building the first base on the planet.[17]

There are several technical obstacles SpaceX will need to overcome, from combatting radiation to producing oxygen, food, and water, but the first issue that will need to be solved before humans get to Mars will be developing a fueling system on the planet that will allow the craft to make the return journey. The Starship will be put into the Earth's orbit, refueled in space, and then sent on its way to Mars. But to return it will need more fuel, and although there are solid scientific theories and techniques for extracting hydrogen from the resources available on Mars, it's a long way from being tested. Assuming that challenge is completed, more Starships will land on the surface, each bringing people and equipment to build the first city on the planet in the 2030s.

Musk's plan, although ambitious, holds more promise than any other Mars colonization proposal. But it's destined to run headlong into the Outer Space Treaty at some point. Michael Listner, an attorney and founder of the Space Law & Policy Solutions law firm, told *Outline* that the treaty conflicts directly with Musk's plans.[18] "When a private citizen makes a claim to private, real property, basically, that's saying the United States is making that claim as well, because of that continuing jurisdiction that the U.S. government always has," Listner said. "But you can't do that." As space lobbying gains a stronger hold in Washington and the might of large companies like SpaceX increases, Listner likely won't be right about that for long. Humans will land on Mars relatively soon, and whether as insurance against the fragility of Earth, or just to take humans further into the unknown, a colony will follow.

This is difficult to debate and discuss because it's hard for humans to truly wrap their heads around something that seems so far away, both in terms of time and distance. But given the apparent adherence to the "first there makes the rules" policy of space, there are some real dangers for any population when they get there. For one thing, if a corporation is first to establish a colony, the people of Mars are going to have a hard time keeping hold of their freedom.

The idea of monopolies in space get more dangerous the further from Earth they may occur, and they call into question the very concept of freedom in space. When you stray into the future of space travel, you'll inevitably contemplate settlements on either the Moon, Mars, or other planets. But what kind of life could be expected in such an environment, and how does that differ if the mission is led by a corporation rather than a government?

"Sometimes people think about escaping Earth to escape traditions and restrictions on the planet. They think about space as a free environment, and to a certain extent that's true, but the problem in space is that every commodity you need to survive, whether that's water, food, or particularly oxygen that you need on a second-to-second timescale, has to come through a manufacturing process," said Charles Cockell, professor of astrobiology at the University of Edinburgh.[19] "By default, you've got this great interdependence with other people in a way that is rarely seen on the Earth. Even in the most extreme environments you can still breathe your own oxygen. So you got the potential for tyranny in space that's to an extreme that I don't think has really ever been seen on the Earth."

Cockell has written about this topic at length and put together a collection of views from himself and others on the topic of freedom in space. He thinks it will be extremely difficult to ensure the freedoms of individuals on any colony or settlement on another planet, but there are some measures that can be taken to improve the situation. He says it will be imperative that monopolies must be minimized, giving the example of one company controlling the oxygen supply of a colony. Aside from doing so legally, this could also be done through engineering.

"If you think about the way that you can get oxygen on Mars, you can dig up permafrost and you can melt it and then clean it and electrolyze it and that'll produce oxygen. But there are a lot of processes between digging up permafrost, melting it, cleaning it, and electrolyzing it," explained Cockell. "Whereas you can also get oxygen by simply taking carbon dioxide from the Martian atmosphere and then chemically breaking it down into oxygen in a single machine. What you can do is start thinking about what are the best ways of extracting oxygen from any environment to minimize the number of steps between the material that has the oxygen atoms and a breath of air—in other words, minimizing the possibilities for tyranny."

Mars is the location for many excellent science-fiction stories, but perhaps the most prescient has been *The Martian Chronicles* series, the 1950 short story fix-up by Ray Bradbury.[20] In these stories, the people of Earth are obsessed with getting to Mars, and achieve that feat only to find it well defended by the current inhabitants. Earthlings overpower the Martians and colonize Mars, but are unable to leave the worst of humanity back on Earth. Eventually, the human population of Mars rushes back to Earth, eager to help in the aftermath of a nuclear war they've witnessed from afar.

The stories are filled with warnings that should be heeded today, as the prospect of humans on Mars moves ever nearer. Part of the appeal of visiting a new planet is the notion that we can start again as a species. Musk, for example, envisions a direct democracy as the political system of a colony on Mars. That's the kind of clean slate people use Mars to symbolize. The reality is a little more sobering, as many of the characters in *The Martian Chronicles* discovered. Settlers on Mars would rely so heavily on Earth for crucial resources, and they would be so closely linked with whatever country they launched from through the Outer Space Treaty, that they couldn't help but act like earthlings for at least the first few generations that humanity lived there. The species can go to the Red Planet and make it great, an example for Earth to follow. Or we can let the worst of humanity follow us on the nine-month journey, and miss the chance to take the best of our civilization further out into the solar system. If private space companies are destined to take us on this journey, we'll be building our second home on the same shaky foundations of capitalism on which we've built our first home.

Conclusion

HUMANS TEND TO SCATTER THEMSELVES AS FAR AND WIDE AS possible. What started as a species confined to one continent has reached every corner of the world and now looks beyond it. Unfortunately, we've never been content with taking just what we need, so the dispersion of our kind has coincided with the failing health of the planet. Nothing has driven our utter domination of this world more than the gears of capitalism, pushing us to great new heights in technological achievement, but simultaneously embracing our desire to hoard resources.

Consider the world's greatest spacefaring superpower, the country the majority of commercial space companies call home—the United States. The distribution of wealth in the country could hardly be more skewed. People die regularly because they can't afford to go to a hospital or buy health insurance. Inequality rages to the point where the color of a person's skin in America drastically effects the chances they will die young or go to prison. "Patriots" preach their right to own a machine gun as children in schools are gunned down by the same weapons with depressing regularity. Meanwhile, the government, infected by the money of lobbying groups, is unable to break its dependence on energy sources that are destroying the planet.

America isn't the only country with these problems; it's just the best example. And one thing links all of our woes—money. Every single day in America, companies make billions of dollars in profits from people suffering from cancer, young adults going to prison for minor crimes, fear-ridden fools with an addiction to firepower, and energy sources that make our air unbreathable and our futures uncertain. Capitalism fuels innovation, yes, but left unchecked it comes at a heavy cost.

Space, despite being a cold, harsh environment full of danger, still represents something of a blank slate for humanity. For many, among the mountains of Mars lies the potential for a fresh start, a utopia that shrugs off the mistakes of our history and forges the species ahead into a new age of exploration, peace, and prosperity for all. But to sustain any hope of this happening, we need to temper the instincts that have brought us to the point where we need to establish a new home before we destroy this one. That means finding a way for commercial interests and the good of humanity to work side by side.

As we've seen in the chapters of this book, we're not off to the best start. Companies have thrown satellites into orbit with such reckless abandon it's only a matter of time before the cluttered avenues surrounding our planet become a rampaging, ever-growing cloud of debris. Money raised to fund these adventures has already found its way into the pockets of politicians, buying influence for single companies rather than noble causes. Space tourism, when it restarts, will still only be an option for the uber-wealthy. The threat of monopolies in space lingers over the entire commercial space industry, and nobody seems to have given it a second thought. We look set to head into murky waters in the quest to make it safe for humans to travel long distances in the universe. And when companies reach the Moon and other planets, they plan to sidestep the most important treaty attempting to protect our interests in space. At every turn there is inequality, exclusivity, and potential for exploitation.

But we still have a chance for the partnerships and regulations we need to make space travel sustainable. Some of the figures I've met in writing this book give me real hope, from the relentless space lawyers to the well-intentioned investors and CEOs. Some founders and entrepreneurs in the areas of commercial space show a sense of responsibility to do something good for the world, not just make money. None of the companies I've spoken to in the commercial space sector harbor anything approaching malicious intent, but they are very much the pioneers, and the industry is still small enough to be policed. The unfortunate irony of commercial expansion is that regulations will likely drop off as more companies enter the sector, meaning more players abiding by fewer rules.

Great things have already come from commercial space companies, the most important being reusable rockets and the lowering of launch costs. This is a significant step toward the democratization of space, and a sign that the goals of a company can align with those of the world. The goals of the major players, like Elon Musk, appear to be well intentioned and true. Musk has risked a great deal in his pursuit and, despite his eccentricities, appears to truly want to get to Mars for the good of humanity, rather than his bank balance. We should insist that the industries emerging off the back of his efforts share the same desire and put measures in place to ensure it. The intentions of Jeff Bezos to save the planet by taking the world's manufacturing and industry into space are also apparently well-meaning, but the idea of such a powerful capitalist controlling a vast off-world economy is—and should be—terrifying.

It's a difficult campaign to make the world pay attention to a far-off danger, particularly when there are plenty of problems in the present to worry about. But to ignore the possibility of runaway capitalism in space would be dangerous, and tough to reverse.

We will hit several major milestones in the coming years, as the pace of change quickens. The launch of the new space tourism will most likely come first, with Virgin Galactic racing toward the finish line and hoping to take its founder Richard Branson to space in 2019 (if they succeed, they could beat this book to publication). Around the same time, SpaceX will take astronauts up to the International Space Station, marking the first instance a commercial rocket has launched passengers into space. These two events will legitimize private companies further and, after that, most likely spur a sharp uptick in achievements.

Perhaps humanity is most likely to reproduce its bad behaviors on the Moon, where a good, old-fashioned resource race may break out at some point next decade. And the space station race is also worth monitoring against monopolies. We must hope, too, that there will be constant vigilance over the mess of debris circling the planet.

Every successful venture in space will bring us closer to the ultimate goal—a colony on either Mars or the Moon. Predicting the future is a fool's game, but there are two standout potential outcomes to the inevitable lunar or Martian experiment. Either it follows the path of the early International Space Station, a landmark of cooperation between countries, companies, and individuals, and acts as a shining beacon for the rest of us on Earth. Or it becomes a mash-up of corporate interests and a repetition of generations of mistakes.

The Outer Space Treaty was a great achievement in its time, but will soon find itself out of date, and needs to be revisited and revised specifically for these times of commercial interests in space. Without strong revisions, the document's vague wording will be its downfall, and the gigantic loopholes will be exploited. When that conversation does happen, private companies should be involved, not through bought-and-paid-for politicians, but with a transparent and open seat at the table. Perhaps we can use this as an opportunity to keep fractious nations in check as well, banning all off-world conflict entirely and reinforcing the cooperative spirit needed in space.

Each country must also put their own restrictions in place, as the onus of responsible space activities will likely always fall mostly on the shoulders of individual governments. The temptation to attract the wealth of the space industry cannot be allowed to cause a regulatory race to the bottom. For this, we need politicians to look to the future and act in the best interest of the species, rather than their next campaign funding—a tough task in some countries, including our own. But that's exactly what we need before we can sit back and enjoy, or even one day participate in, the revolution underway in space.

Acknowledgments

WRITING A BOOK ABOUT A RAPIDLY CHANGING TOPIC LIKE space can feel like landing a rover on an asteroid speeding through the cosmos. In that metaphor, I would never have left the launch pad if it weren't for the support of those around me. It's important I acknowledge my parents and family for raising me in an environment where I was afforded the luxury of gazing into the stars and dreaming of the wider universe. Without the curiosity and eagerness to learn they instilled in me, I would not have had the inspiration to write this or any book.

I've had many great professors, editors, and mentors who have helped shaped my writing and thinking, and I'm eternally grateful to all of them. There is one editor I will mention by name, Chris Mohney, who commissioned the original story on the benefits of a mission to Mars, which inspired me to delve deeper into space. And it's easy to take space exploration for granted, so I would like to pay my respects to the momentous, heroic, and awe-inspiring work of any company, country, space agency or individual that has ventured into the unknown with the goal of bettering humanity.

While I have spent my entire professional career writing as a journalist, this was my first book, and I would never have undertaken such a huge task if it weren't for the prodding and prompting of Michael Barron, who filled me with confidence and gave me the kind of guidance a first-time author could only dream of. Even the proposal writing process of this book was fairly hectic, such was my desire to get the book into the world before the commercial space industry zoomed straight by. Luckily, I could rely on feedback from colleagues and friends. Sean Williams, my long-time collaborator, was a great help at several crucial points in the writing, and a band of former colleagues—led by Esme Benjamin and Amber Snider—were always happy to read chapters at varying degrees of completion. I absolutely need to thank my editor Ryan Harrington, whose efficiency, insight, and patience was all so greatly appreciated. If my writing got us into an asteroid's general vicinity, Ryan was the one who deftly landed it.

The most important acknowledgment should go to the person closest to me, who has been with me every step of the way. My wife Seren inspires me to make the best of myself, gives me the confidence to go out into the world, and supports me in every single way. I couldn't have written a word without her.

Notes

INTRODUCTION

1 "First Moon Landing Fast Facts," CNN, July 13, 2018, https://www.cnn.com/2013/09/15/us/moon-landing-fast-facts/index.html.

2 David S. F. Portree, "Ending Apollo (1968)," *Wired*, June 3, 2017, https://www.wired.com/2013/09/ending-apollo-1968/.

3 Robert Z. Pearlman, "NASA's Space Shuttle Program Officially Ends After Final Celebration," Space.com, September 1, 2011, https://www.space.com/12804-nasa-space-shuttle-program-officially-ends.html.

4 Olivia Solon. "Elon Musk: We Must Colonise Mars to Preserve Our Species in a Third World War." The Guardian. March 11, 2018. Accessed May 09, 2019. https://www.theguardian.com/technology/2018/mar/11/elon-musk-colonise-mars-third-world-war.

5 Alan Boyle, "Life, Liberty and the Pursuit of Spaceflight? Jeff Bezos Links Blue Origin to Saving Earth," *GeekWire*, November 16, 2017, https://www.geekwire.com/2017/life-liberty-pursuit-spaceflight-jeff-bezos-links-blue-origin-saving-earth/.

6 Editors, "Transcontinental Railroad," History.com, April 20, 2010, https://www.history.com/topics/inventions/transcontinental-railroad.

7 Author interview with Dr. James Vedda, December 10, 2018.

8 "Landmark Legislation: The Pacific Railway Act of 1862," U.S. Senate: Landmark Legislation, January 12, 2017, https://www .senate.gov/artandhistory/history/common/generic/PacificRail wayActof1862.htm.

CHAPTER 1

1 Brian Dunbar, "The Apollo-Soyuz Mission," NASA, April 16, 2015, https://www.nasa.gov/mission_pages/apollo-soyuz/astp _mission.html.

2 "Ford Hails 2 Launchings as a Mark of Cooperation," *New York Times*, July 16, 1975, https://www.nytimes.com/1975/07/16 /archives/ford-hails-2-launchings-as-a-mark-of-cooperation.html.

3 Brian Dunbar, "Remembering the Legacy of President Gerald Ford," NASA, https://www.nasa.gov/vision/space/features/ger- ald_ford.html (accessed March 29, 2019).

4 Melanie Whiting, "Skylab: America's First Space Station," NASA, May 14, 2018, https://www.nasa.gov/feature/skylab-america -s-first-space-station.

5 Collier's Articles on the Conquest of Space (1952–1954), Space Stuff, https://www.rmastri.it/spacestuff/wernher-von-braun /colliers-articles-on-the-conquest-of-space-1952-1954/ (accessed March 29, 2019).

6 "Korolev and Freedom of Space: February 14, 1955–October 4, 1957," NASA, https://history.nasa.gov/monograph10/korspace. html (accessed March 29, 2019).

7 "Eisenhower Presents His 'Open Skies' Plan," History.com, November 13, 2009, https://www.history.com/this-day-in-history/ eisenhower-presents-his-open-skies-plan.

8 Elizabeth Howell, "Sputnik: The Space Race's Opening Shot," Space.com, August 22, 2018, https://www.space.com/17563-spu nik.html.

9 "National Affairs: PROJECT VANGUARD," *Time*, October 21, 1957, http://content.time.com/time/magazine/article/0,9171,937 919,00.html.

10 Brian Dunbar, "The Birth of NASA," NASA, March 28, 2008, https://www.nasa.gov/exploration/whyweexplore/Why_We _29.html.

11 "50th COPUOS," International Astronautical Federation, http:// www.iafastro.org/events/50th-copuos/ (accessed February 5, 2019).

12 Treaty on Principles Governing the Activities of States in the Exploration and Use of Outer Space, Including the Moon and Other Celestial Bodies, U.S.-U.K.-U.S.S.R., article IV, UN Office of Disarmament Affairs, opened for signature January 27, 1967, resolution 2222 (XXI), http://disarmament.un.org /treaties/t/outer_space/text.

13 Lyndon B. Johnson, "Remarks at the Signing of the Treaty on Outer Space, January 27, 1967," in *Public Papers of the Presidents of the United States: Lyndon B. Johnson, 1967*, vol. I, pp. 91–92 (Washington, D.C.: Government Printing Office, 1968), available from LBJ Presidential Library. http://www.lbjlibrary.net/collections/ selected-speeches/1967/01-27-1967.html.

14 Treaty on Principles Governing the Activities of States in the Exploration and Use of Outer Space, article II.

15 "ESA Remembers the Night of the Comet," ESA Science & Technology, http://sci.esa.int/giotto/48529-esa-remembers-the-night-of-the-comet/ (accessed January 5, 2019).

16 "Towards Competition in International Satellite Services: Rethinking the Role of INTELSAT," National Archives and Records Administration, https://clintonwhitehouse4.archives.gov/ WH/EOP/CEA/html/paper.html (accessed December 14, 2018).

17 Carl White, "Greenwich Maverick Rene Anselmo," Greenwich Library, May 12, 2016, https://www.greenwichlibrary.org/ greenwich-maverick-rene-anselmo-2/.

18 David Cay Johnston, "Rene Anselmo, 69, the Founder of a Satellite Network, Is Dead," *New York Times*, September 21, 1995, https://www.nytimes.com/1995/09/21/obituaries/rene-anselmo-69-the-founder-of-a-satellite-network-is-dead.html.

19 Author interview with Jeffrey Manber, CEO of NanoRacks, December 19, 2018.

20 Alan Breznick, "The Little Satellite Company That Could," Broadcasting & Cable, September 27, 2004, https://www.broadcastingcable.com/news/little-satellite-company-could-74059.

21 "History of PanAmSat Corporation," FundingUniverse, http:// www.fundinguniverse.com/company-histories/panamsat-corpora tion-history/ (accessed January 12, 2019).

22 Johnston, "Rene Anselmo, 69, the Founder of a Satellite Network, Is Dead."

CHAPTER 2

1 "Mir Space Station," NASA, https://history.nasa.gov/SP-4225/mir/mir.htm (accessed December 17, 2018).
2 Frank Slazer, "Cooperation in Space: The International Space Station Benefits from ISO Standards," ISO, October 5, 2011, https://www.iso.org/news/2011/10/Ref1555.html.
3 Author interview with Jeffery Manber, CEO, NanoRacks, December 19, 2018.
4 Ibid.
5 "Mir Space Station," NASA.
6 Author interview with Bill Stone, chairman and CTO, Shackleton Energy, October 11, 2018.
7 Ibid.
8 Brian Dunbar, "Clementine Mission," NASA, https://www.nasa.gov/mission_pages/LCROSS/searchforwater/clementine.html (accessed December 18, 2018).
9 Alan Boyle, "Russia Bids Farewell to Mir," NBC News, June 27, 2006, http://www.nbcnews.com/id/3077781/ns/technology_and_science-space/t/russia-bids-farewell-mir/#.XJ6FZ5hKg2.

CHAPTER 3

1 Author interview with Esther Dyson, investor, October 7, 2018.
2 Mike Wall, "First Space Tourist: How a U.S. Millionaire Bought a Ticket to Orbit," Space.com, April 27, 2011, https://www.space.com/11492-space-tourism-pioneer-dennis-tito.html.
3 Greg Klerkx, *Lost in Space: The Fall of NASA and the Dream of a New Space Age* (New York: Vintage Books, 2005).
4 Klerkx, *Lost in Space*.
5 Neil Strauss, "Elon Musk: The Architect of Tomorrow," *Rolling Stone*, June 25, 2018, https://www.rollingstone.com/culture/culture-features/elon-musk-the-architect-of-tomorrow-120850/.
6 Author interview with Dr. Robert Zubrin, founder, Mars Society, January 14, 2019.

7 Mark Kaufman, "The Bizarre Story of When SpaceX Tried to Buy Missiles from Russia," Inverse, August 2, 2017, https://www.inverse.com/article/34976-spacex-ceo-elon-musk-tried-to-buy-icbm-rockets-from-russia.

8 Andrew Chaikin, "Is SpaceX Changing the Rocket Equation?" *Air & Space Magazine*, January 1, 2012, https://www.airspacemag.com/space/is-spacex-changing-the-rocket-equation-132285884/.

9 Author interview with Stephen Attenborough, commercial director, Virgin Galactic, October 2, 2018.

10 Leonard David, "SpaceShipOne Wins $10 Million Ansari X Prize in Historic 2nd Trip to Space," Space.com, October 4, 2004, https://www.space.com/403-spaceshipone-wins-10-million-ansari-prize-historic-2nd-trip-space.html.

11 Author interview with Esther Dyson.

CHAPTER 4

1 William Langewiesche, "Columbia's Last Flight," *The Atlantic*, November 2003, https://www.theatlantic.com/magazine/archive/2003/11/columbias-last-flight/304204/.

2 "NASA Commercial Orbital Transportation Services," NASA, https://www.nasa.gov/offices/c3po/about/c3po.html (accessed January 25, 2019).

3 "President Bush Offers New Vision for NASA," NASA, January 14, 2004, https://www.nasa.gov/missions/solarsystem/bush_vision.html.

4 Ashlee Vance, "Elon Musk's Space Dream Almost Killed Tesla," *Bloomberg*, May 14, 2015, https://www.bloomberg.com/graphics/2015-elon-musk-spacex/.

5 Vance, "Elon Musk's Space Dream Almost Killed Tesla."

6 Alexis Madrigal, "Obama Gives NASA More Money, Cuts Manned Trip to Moon," *Wired*, June 4, 2017, https://www.wired.com/2010/02/nasa-budget/.

7 Tariq Malik, "President Obama Signs New Vision for U.S. Space Exploration into Law," Space.com, October 11, 2010, https://www.space.com/9305-president-obama-signs-vision-space-exploration-law.html.

8 Dave Mosher, "NASA 'Will Eventually Retire' Its New Mega-Rocket if SpaceX, Blue Origin Can Safely Launch Their Own Powerful Rockets," *Business Insider*, November 19, 2018, https://www.businessinsider.com/nasa-sls-replacement-spacex-bfr-blue-origin-new-glenn-2018-11.

9 Jeff Foust, "NASA Inspector General Sharply Criticizes SLS Core Stage Development," SpaceNews.com, October 10, 2018, https://spacenews.com/nasa-inspector-general-sharply-criticizes-sls-core-stage-development/.

10 Author interview with Dallas Bienhoff, founder, Cislunar Space Development Company, December 18, 2018.

11 John Noble Wilford, "Reagan Is Reported Near Decision to Approve a New Space Shuttle," *New York Times*, May 25, 1986, https://www.nytimes.com/1986/05/25/us/reagan-is-reported-near-decision-to-approve-a-new-space-shuttle.html.

12 Peter Pae, "Boeing to Pay Fine of $615 Million," *Los Angeles Times*, May 16, 2006, https://www.latimes.com/archives/la-xpm-2006-may-16-fi-boeing16-story.html.

13 Brian Berger and Warren Ferster, "Boeing-Lockheed Rocket Merger Sparks Concern, Legal Challenge," Space.com, October 24, 2005, https://www.space.com/1709-boeing-lockheed-rocket-merger-sparks-concern-legal-challenge.html.

14 Author interview with Barry Lynn, executive director, Open Markets Institute, January 10, 2019.

15 Berger and Ferster, "Boeing-Lockheed Rocket Merger Sparks Concern, Legal Challenge."

16 Eric Berger, "How America's Two Greatest Rocket Companies Battled from the Beginning," *Ars Technica*, August 2, 2017, https://arstechnica.com/science/2017/08/how-americas-two-greatest-rocket-companies-battled-from-the-beginning/.

17 Ibid.

18 Ibid.

19 Matthew Nussbaum, "How Elon Musk Exposed Billions in Questionable Pentagon Spending," *Politico*, May 15, 2016, https://www.politico.com/story/2016/05/elon-musk-rocket-defense-223161.

20 Ibid.

21 Berger "America's Two Greatest Rocket Companies Battled from the Beginning."

22 Sandra Erwin, "Air Force Awards Launch Vehicle Development Contracts to Blue Origin, Northrop Grumman, ULA," SpaceNews.com, January 29, 2019, https://spacenews.com/air-force-awards-launch-vehicle-development-contracts-to-blue-origin-northrop-grumman-ula/.

23 Author interview with Bill Stone, chairman and CTO, Shackleton Energy, October 11, 2018.

CHAPTER 5

1 Author interview with Chad Anderson, CEO, Space Angels, December 20, 2018.

2 Space Angels, *Space Investment Quarterly: Q4 2018*, https://www.spaceangels.com/post/space-investment-quarterly-q4-2018.

3 Paul Ceruzzi, "Apollo Guidance Computer and the First Silicon Chips," National Air and Space Museum, March 22, 2017, https://airandspace.si.edu/stories/editorial/apollo-guidance-computer-and-first-silicon-chips.

4 Scott Rosenberg, "Silicon Valley's First Founder Was Its Worst," *Wired*, July 21, 2017, https://www.wired.com/story/silicon-valleys-first-founder-was-its-worst/.

5 Author interview with Sidney Nakahado, CEO and cofounder, New York Space Alliance, December 10, 2017.

6 Dave Levinthal, "Space New Frontier for Tech Lobbying," *Politico*, July 29, 2011, https://www.politico.com/story/2011/07/space-new-frontier-for-tech-lobbying-060259.

7 Eric Berger, "A Shadowy Op-ed Campaign Is Now Smearing SpaceX in Space Cities," *Ars Technica*, October 4, 2018, https://arstechnica.com/science/2018/10/a-shadowy-op-ed-campaign-is-now-smearing-spacex-in-space-cities/.

8 Dave Mosher, "Boeing May Have Used a Lobbying Firm to Plant a Scathing Opinion Piece About SpaceX in US News Outlets. At Stake Are Billions of Dollars in NASA Contracts," *Business Insider*, October 7, 2018, https://www.businessinsider.com/boeing-may-have-used-firm-to-plant-anti-spacex-oped-2018-10.

9 Author interview with Chris Carberry, president, Explore Mars, December 21, 2018.

10 David S. Cloud and Noah Bierman, "Trump Backed 'Space Force' After Months of Lobbying by Officials with Ties to Aerospace Industry," *Los Angeles Times*, August 18, 2018, https://www.latimes.com/nation/la-na-pol-trump-space-20180817-story.html.

11 "Rep. Jim Cooper—Campaign Finance Summary," OpenSecrets, https://www.opensecrets.org/members-of-congress/summary?cid=N00003132&cycle=2018 (accessed December 3, 2018).

12 Author interview with Eric Stallmer, president, Commercial Space Federation, December 21, 2018.

13 Sarah Kaplan, "President Trump Relaunches the National Space Council," *Washington Post*, June 30, 2017, https://www.washingtonpost.com/news/speaking-of-science/wp/2017/06/30/trump-relaunches-the-national-space-council/?utm_term=.79f56797cddb.

CHAPTER 6

1 Associated Press, "Lance Bass Officially Kicked Off Space Flight," *Billboard*, January 4, 2013, https://www.billboard.com/articles/news/74272/lance-bass-officially-kicked-off-space-flight.

2 Leonard David, "Explosion Kills Three at Mojave Air and Space Port," Space.com, July 27, 2007, https://www.space.com/4123-explosion-kills-mojave-air-space-port.html.

3 Interview with Adam Cohen, Virgin Galactic customer, September 17, 2018.

4 Tariq Malik, "Deadly SpaceShipTwo Crash Caused by Co-Pilot Error: NTSB," Space.com, July 28, 2015, https://www.space.com/30073-virgin-galactic-spaceshiptwo-crash-pilot-error.html.

5 "Branson's Virgin Galactic Reaches Edge of Space," BBC News, December 13, 2018, https://www.bbc.com/news/business-46550862.

6 Eric M. Johnson, "Exclusive: Jeff Bezos Plans to Charge at Least $200,000 for Space . . ." Reuters, July 13, 2018, https://www.reuters.com/article/us-space-blueorigin-exclusive/exclusive-jeff-bezos-plans-to-charge-at-least-200000-for-space-rides-sources-idUSKBN1K301R.

7 Stefanie Waldek, "How to Become a Space Tourist: 8 Companies (Almost) Ready to Launch," *Popular Science*, April 20, 2018, https://www.popsci.com/how-to-become-a-space-tourist.

8 Jeff Foust, "A Short History of Lunar Space Tourism," Space News.com, March 27, 2017, https://spacenews.com/a-short-history-of-lunar-space-tourism/.

9 Sam Levin and Justin McCurry, "Elon Musk to Launch Japanese Billionaire on Space X Rocket to the Moon," *The Guardian*, September 18, 2018, https://www.theguardian.com/technology/2018/sep/17/elon-musk-space-x-moon-rocket-japanese-billionaire.

10 Author interview with Stephen Attenborough, commercial director, Virgin Galactic, October 2, 2018.

11 Frank White, *The Overview Effect: Space Exploration and Human Evolution* (Reston, VA: American Institute of Aeronautics and Astronautics, 2014).

12 Denise Lineberry, "Celebrate with the World for One Night: Yuri's Night," NASA, January 6, 2019, https://www.nasa.gov/centers/langley/news/researchernews/rn_yurisnight.html.

13 White, The Overview Effect.

14 Peter Ward, "The Finnish Entrepreneur Who Wants to Take the World to Space," Culture Trip, February 15, 2018, https://theculturetrip.com/europe/finland/articles/the-finnish-entrepreneur-who-wants-to-take-the-world-to-space/.

15 Sofia Virtanen, "Suomalaistaustainen Avaruusyhtiö Space Nation Konkurssiin—Applessa Hankittu Markkinointikokemuskaan Ei Auttanut: Peter Vesterbacka, Saku Koivu Ja Yli 500 Muuta Menettivät Sijoituksensa," T&T, November 20, 2018, https://www.tekniikkatalous.fi/tiede/avaruus/suomalaistaustainen-avaruusyhtio-space-nation-konkurssiin-applessa-hankittu-markkinointikokemuskaan-ei-auttanut-peter-vesterbacka-saku-koivu-ja-yli-500-muuta-menettivat-sijoituksensa-6749925.

16 Author interview with Hjörtur Smárason via Facebook Messenger, March 18, 2019

CHAPTER 7

1 "Space Debris by the Numbers," European Space Agency. https://www.esa.int/Our_Activities/Operations/Space_Safety_Security/

Space_Debris/Space_debris_by_the_numbers (accessed January 23, 2019).

2 Leonard David, "Effects of Worst Satellite Breakups in History Still Felt Today," Space.com, January 28, 2013, https://www.space.com/19450-space-junk-worst-events-anniversaries.html.

3 Mark Garcia, "Space Debris and Human Spacecraft," NASA, April 14, 2015, https://www.nasa.gov/mission_pages/station/news/orbital_debris.html.

4 Author interview with Moriba Jah, associate professor, University of Texas at Austin, November 28, 2018.

5 Forecast International, "The Market for Civil & Commercial Remote Sensing Satellites," July 2018, https://www.globenewswire.com/news-release/2018/07/17/1538020/0/en/Remote-Sensing-Satellite-Market-Takes-Off-Driven-by-SmallSat-Deliveries.html.

6 "SpaceX Gets Nod to Put 12,000 Satellites in Orbit," Phys.org, November 16, 2018, https://phys.org/news/2018-11-spacex-satellites-orbit.html.

7 Tara Halt and Anna Wieger, *Smallsats by the Numbers 2019* (Washington, D.C.: Bryce Space and Technology, 2019).

8 Loren Grush. "As Satellite Constellations Grow Larger, NASA Is Worried about Orbital Debris." The Verge. September 28, 2018. https://www.theverge.com/2018/9/28/17906158/nasa-spacex-oneweb-satellite-large-constellations-orbital-debris. Accessed December 22, 2018.

9 Jonathan Amos, "RemoveDebris: UK Satellite Nets 'Space Junk,'" BBC News, September 19, 2018, https://www.bbc.com/news/science-environment-45565815.

10 Author interview with Jack Wengryniuk, vice president of regulatory engineering, Inmarsat. January 10, 2019.

11 Author interview with Yousaf Butt, astrophysicist and consultant on international space policy, November 30, 2018.

12 Marina Koren, "Launching Rogue Satellites into Space Was a 'Mistake,'" *The Atlantic*, September 7, 2018, https:/www.theatlantic.com/technology/archive/2018/09/spacebees-swarm-unauthorized-satellite-launch/569395/.

13 Mark Harris, "FCC Accuses Stealthy Startup of Launching Rogue Satellites," IEEE Spectrum: Technology, Engineering, and

Science News, March 9, 2018, https://spectrum.ieee.org/tech-talk/aerospace/satellites/fcc-accuses-stealthy-startup-of-launching-rogue-satellites.

14 Michelle La Vone, "The Kessler Syndrome Explained," *Space Safety* magazine, http://www.spacesafetymagazine.com/space-debris/kessler-syndrome/ (accessed November 24, 2018).

CHAPTER 8

1 Carin Zissis, "China's Anti-Satellite Test," Council on Foreign Relations, February 22, 2007, https://www.cfr.org/backgrounder/chinas-anti-satellite-test.

2 "Space Debris by the Numbers," European Space Agency, https://www.esa.int/Our_Activities/Operations/Space_Safety_Security/Space_Debris/Space_debris_by_the_numbers (accessed January 24, 2019).

3 Michael Greshko, "China Just Landed on the Far Side of the Moon: What Comes Next?" *National Geographic*, January 3, 2019, https://www.nationalgeographic.com/science/2019/01/china-change-4-historic-landing-moon-far-side-explained/.

4 Liangyu, "China's Chang'e-4 Probe Soft-Lands on Moon's Far Side," Xinhua, January 3, 2019, http://www.xinhuanet.com/english/2019-01/03/c_137716998.htm.

5 Kerry Brown, "Qian Xuesen Obituary," *The Guardian*, November 1, 2009, https://www.theguardian.com/technology/2009/nov/01/qian-xuesen-obituary.

6 Evan Osnos, "The Two Lives of Qian Xuesen," *New Yorker*, June 18, 2017, https://www.newyorker.com/news/evan-osnos/the-two-lives-of-qian-xuesen.

7 Author interview with Brian Harvey, space analyst, January 8, 2019.

8 "A Brief History of China in Space," *The Telegraph*, August 24, 2011, https://www.telegraph.co.uk/news/science/space/8719848/A-brief-history-of-China-in-space.html.

9 Alex Stuckey, "Why Chinese Astronauts Are Banned from the International Space Station, NASA Activities," *Houston Chronicle*, February 22, 2018, https://www.chron.com/news/nation-world/

space/article/Ever-wonder-why-you-ve-never-seen-a-Chinese-12631096.php.

10 Nina Xiang, "Chinese VCs Are Hunting for China's SpaceX, But It May Still Be Years Away," *Forbes*, August 24, 2018, https://www.forbes.com/sites/ninaxiang/2018/08/23/chinese-vcs-are-hunting-for-chinas-spacex-but-it-may-still-be-years-away/#3475005867f1.

11 Sally French, "China Has 9 of the World's 20 Biggest Tech Companies," MarketWatch, May 31, 2018, https://www.marketwatch.com/story/china-has-9-of-the-worlds-20-biggest-tech-companies-2018-05-31.

12 Li Yan, "Commercial Space Sector Poised for Takeoff," China News Service, January 7, 2019, http://www.ecns.cn/news/sci-tech/2019-01-07/detail-ifzcitha9951844.shtml.

13 Marina Koren, "China's Growing Ambitions in Space," *The Atlantic*, January 24, 2017, https://www.theatlantic.com/science/archive/2017/01/china-space/497846/.

CHAPTER 9

1 "What Is the ISS National Lab?" NASA, March 31, 2015, https://www.nasa.gov/mission_pages/station/research/nlab/what_is_nlab.

2 Loren Grush, "Trump Administration Wants to End NASA Funding for the International Space Station by 2025," The Verge, January 25, 2018, https://www.theverge.com/2018/1/24/16930154/nasa-international-space-station-president-trump-budget-request-2025.

3 Author interview with Amir Blachman, chief business officer, Axiom Space, December 13, 2018.

4 David M. Ewalt, "Cosmic Landlord," Forbes, June 8, 2011, https://www.forbes.com/forbes/2011/0627/features-robert-bigelow-aerospace-real-estate-cosmic-landlord.html#1e198f747712.

5 "TransHab Concept," NASA, https://spaceflight.nasa.gov/history/station/transhab/ (accessed January 12, 2019).

6 Ryan Bradley, "Can Billionaire Robert Bigelow Create a Life for Humans in Space?" *Popular Science*, April 8, 2016, https://www.

popsci.com/can-billionaire-robert-bigelow-create-a-life-for-humans-in-space.

7 Mike Wall, "Private Inflatable Habitat Will Stay at Space Station for at Least 3 More Years," Space.com, December 5, 2017, https://www.space.com/38983-nasa-extends-beam-inflatable-habitat-space-station.html.

8 "BEAM," Bigelow Aerospace, http://bigelowaerospace.com/pages/beam/ (accessed January 12, 2019).

9 "B330," Bigelow Aerospace, http://bigelowaerospace.com/pages/b330/ (accessed January 12, 2019).

10 Dave Mosher, "A New Company Plans to Launch Huge, Inflatable Spacecraft into Orbit—and Sell Reservations to Countries and Tourists," *Business Insider*, February 21, 2018, https://www.business-insider.com/private-space-station-hotels-robert-bigelow-2018-2.

11 Author interview with Frank Bunger, CEO, Orion Span, November 21, 2018.

12 Alan Boyle, "One Small Step: Orion Span Kicks off $2M Crowd-funding Campaign for Space Hotel," GeekWire, December 9, 2018, https://www.geekwire.com/2018/one-small-step-orion-span-kicks-off-crowdfunding-campaign-space-hotel/.

13 Author interview with Jeffrey Manber, CEO, NanoRacks, December 19, 2018.

14 Sarah Lewin, "Making Stuff in Space: Off-Earth Manufacturing Is Just Getting Started," Space.com, May 11, 2018, https://www.space.com/40552-space-based-manufacturing-just-getting-started.html.

15 Haylie Kasap. "Exotic Glass Fibers from Space," Upward, December 11, 2018, https://upward.issnationallab.org/the-race-to-manufacture-zblan/.

16 Author interview with Dr. Dimitry Starodubov via email, December 14, 2018.

17 Author interview with Dr. James Vedda, senior policy analyst, Center for Space Policy and Strategy at the Aerospace Corporation, December 10, 2018.

18 Author interview with Barry Lynn, executive director, Open Markets Institute, January 10, 2019.

CHAPTER 18

1 Marilynn Marchione, "Chinese Researcher Claims First Gene-Edited Babies," Associated Press, November 26, 2018, https://www.apnews.com/4997bb7aa36c45449b488e19ac83e86d.

2 Pheobe Zhang, "China Confirms Gene-Edited Babies; Scientist and Staff to Face Punishment," *South China Morning Post*, February 4, 2019, https://www.scmp.com/news/china/science/article/2182964/china-confirms-gene-edited-babies-blames-scientist-he-jiankui.

3 Brad Plumer, Eliza Barclay, Julia Belluz, and Umair Irfan, "A Simple Guide to CRISPR, One of the Biggest Science Stories of the Decade," Vox, December 27, 2018, https://www.vox.com/2018/7/23/17594864/crispr-cas9-gene-editing.

4 Plumer, Barclay, Belluz, and Irfan, "A Simple Guide to CRISPR."

5 Dylan Matthews, "A Genetically Modified Organism Could End Malaria and Save Millions of Lives—if We Decide to Use It," Vox, September 26, 2018, https://www.vox.com/science-and-health/2018/5/31/17344406/crispr-mosquito-malaria-gene-drive-editing-target-africa-regulation-gmo.

6 George Dvorsky, "Chinese Government Says It Has Shut Down Controversial Human Gene-Editing Project," Gizmodo, November 29, 2018, https://gizmodo.com/chinese-government-says-it-has-shut-down-controversial-1830751956.

7 Jennifer Doudna, "CRISPR Co-Inventor Responds to Claim of First Genetically Edited Babies," *Berkeley News*, January 18, 2019, https://news.berkeley.edu/2018/11/26/doudna-responds-to-claim-of-first-crispr-edited-babies/.

8 Jon Cohen, "'I Feel an Obligation to Be Balanced.' Noted Biologist Comes to Defense of Gene Editing Babies," *Science*, November 29, 2018, https://www.sciencemag.org/news/2018/11/i-feel-obligation-be-balanced-noted-biologist-comes-defense-gene-editing-babies.

9 William Herkewitz, "The Secret of the Only Animal That Can Survive in Space," *Popular Mechanics*, February 15, 2018, https://www.popularmechanics.com/space/a11137/secrets-of-the-water-bear-the-only-animal-that-can-survive-in-space-17069978/.

10 Antonio Regalado, "The Not-So-Secret Plan to Genetically Modify Astronauts," *MIT Technology Review*, May 10, 2017, https://www.technologyreview.com/s/604142/engineering-the-perfect-astronaut/.

11 "The Human Body in Space," NASA, March 30, 2016, https://www.nasa.gov/hrp/bodyinspace.

12 Author interview with Dr. George Church, professor of genetics, Harvard Medical School, December 20, 2018.

13 Author interview with Dr. Nick Kanas, emeritus professor of psychiatry, University of California, January 7, 2019.

14 "Sickening Solar Flares," NASA, November 8, 2005, https://www.nasa.gov/mission_pages/stereo/news/stereo_astronauts.html.

15 Sarah Frazier, "How to Protect Astronauts from Space Radiation on Mars," NASA, September 30, 2015, https://www.nasa.gov/feature/goddard/real-martians-how-to-protect-astronauts-from-space-radiation-on-mars.

16 Leonard David, "Anti-Radiation Vest to Get Deep-Space Test Next Year," Space.com, May 15, 2018, https://www.space.com/40590-anti-radiation-life-vest-deep-space.html.

17 Brian Wang, "2 Tesla Magnetic Shield Placed at Mars Lagrange Point Would Shield Martian Atmosphere for Affordable Partial Terraforming," NextBigFuture.com, October 15, 2017, https://www.nextbigfuture.com/2017/03/2-tesla-magnetic-shield-placed-at-mars.html.

18 "Human Genetic Engineering," AP-NORC Center for Public Affairs Research, http://apnorc.org/projects/Pages/Human-Genetic-Engineering.aspx.

19 Bill Gates, "What I Learned at Work This Year," Gates Notes, December 29, 2018, https://www.gatesnotes.com/About-Bill-Gates/Year-in-Review-2018.

CHAPTER 11

1 Author interview with Takeshi Hakamada, CEO, iSpace, June 5, 2018.

2 John Schwartz, "Google Backs $25 Million 'Lunar X Prize,'"

New York Times, September 13, 2007, https://www.nytimes.
com/2007/09/13/technology/13cnd-xprize.html.

3 Author interview with Takeshi Hakamada.

4 Mike Wall, "Ex-Prize: Google's $30 Million Moon Race Ends
with No Winner," Space.com, January 23, 2018, https://www.
space.com/39467-google-lunar-xprize-moon-race-ends.html.

5 Author interview with Takeshi Hakamada.

6 "Project," ispace, https://ispace-inc.com/project/ (accessed 3
September, 2018).

7 "Redefining Natural Resources," Planetary Resources, https://
www.planetaryresources.com/why-asteroids/ (accessed 3 September, 2018).

8 "NASA Announces New Partnerships for Commercial Lunar
Payload Delivery Services," NASA, November 29, 2018, https://
solarsystem.nasa.gov/news/774/nasa-announces-new-partner-
ships-for-commercial-lunar-payload-delivery-services/.

9 "Reopening the American Frontier: Exploring How the Outer
Space Treaty Will Impact American Commerce and Settlement
in Space," U.S. Senate Committee on Commerce, Science, and
Transportation, May 23, 2017, https://www.commerce.senate.
gov/public/index.cfm/2017/5/reopening-the-american-frontier-
exploring-how-the-outer-space-treaty-will-impact-american-
commerce-and-settlement-in-space.

10 Author interview with Frans von der Dunk, University of
Nebraska-Lincoln, College of Law, April 11, 2018.

11 Brooks Hays, "New U.S. Space Mining Law May Violate International Treaty," United Press International, November 27, 2015,
https://www.upi.com/Science_News/2015/11/27/New-US-
space-mining-law-may-violate-international-treaty/
8751448634436/.

12 Committee on the Peaceful Uses of Outer Space Legal Subcommittee, fifty-seventh session, Vienna, 9–20, April 2018.

13 Agreement Governing the Activities of States on the Moon and
Other Celestial Bodies, United Nations Office for Outer Space
Affairs, resolution 34/68, opened for signature 1979, http://www.

unoosa.org/oosa/en/ourwork/spacelaw/treaties/intromoon-agreement.html.

14 Author interview with Kyle Acierno, May 3, 2018

CHAPTER 12

1 Author interview with Bill Hargenrader, co-host, Red Planet Radio, December 10, 2018.

2 "Mars One Will Settle Men on Mars in 2023," Mars One, May 31, 2012, https://www.mars-one.com/news/press-releases/mars-one-will-settle-men-on-mars-in-2023.

3 "Mars One Plans Suicide Mission to Red Planet for 2023," Fox News, June 24, 2012, https://www.foxnews.com/science/mars-one-plans-suicide-mission-to-red-planet-for-2023.

4 "Mars One Funding Model: Finance and Feasibility," Mars One, https://www.mars-one.com/faq/finance-and-feasibility/what-is-mars-ones-funding-model (accessed January 12, 2019).

5 Author interview with Sue Ann Pien, Mars One candidate, January 9, 2019.

6 Loren Grush, "Mars One Just Delayed Its (Highly Unlikely) Mars Mission—Again," The Verge, December 7, 2016, https://www.theverge.com/2016/12/7/13869856/mars-one-revised-mission-timeline-again-launch-plan-2031.

7 Adam Gabbatt, "Mars One Says 80,000 Have Applied for One-Way Mission to Red Planet," *The Guardian*, May 8, 2013, https://www.theguardian.com/science/2013/may/08/mars-one-applications-mission; Alan Boyle, "One-Way Mars Trip Attracts 165,000 Would-Be Astronauts . . . and Counting," NBC News, August 23, 2013, https://www.nbcnews.com/science/one-way-mars-trip-attracts-165-000-would-be-astronauts-6C10981032.

8 "Mars One: First Private Mars Mission in 2018," Indiegogo, December 10, 2013, https://www.indiegogo.com/projects/mars-one-first-private-mars-mission-in-2018.

9 Rae Paoletta, "Mars One Is a "Money Grab" Where Everyone Loses," Inverse, March 29, 2018, https://www.inverse.com/article/42965-mars-one-is-a-money-grab-where-everyone-loses.

10 Jennifer Chu, "Mars One (and Done?)," MIT News, October 14,

2014, http://news.mit.edu/2014/technical-feasibility
-mars-one-1014.

11 Loren Grush, "Mars One Debates MIT: CEO Bas Lansdorp Still
Doesn't Have a Plan to Reach the Planet," The Verge, August 18,
2015, https://www.theverge.com/2015/8/18/9166697/mars-one-
plan-mit-debate-ceo-bas-lansdorp.

12 Peter Walker, "Shortlisted Mars One Astronaut Says Venture
Has Taken Giant Leap Backwards," *The Guardian*, March 18,
2015, https://www.theguardian.com/science/2015/mar/18/mars-
one-astronaut-joseph-roche-says-dutch-venture-implausible.

13 Bill Chappell, "Mars One Fizzles into Bankruptcy After Promising a New Life in Space," NPR, February 12, 2019, https://www.
npr.org/2019/02/12/693945694/mars-one-fizzles-into-bank-
ruptcy-after-promising-a-new-life-in-space.

14 "Dr. Robert Zubrin (USA)-Advisers-About Mars One," Mars One,
https://www.mars-one.com/about-mars-one/advisers/dr.-robert-
zubrin-usa (accessed January 12, 2019).

15 Author interview with Dr. Robert Zubrin, founder, Mars Society,
January 14, 2019.

16 Jeff Foust, "Musk Unveils Revised Version of Giant Interplanetary
Launch System," SpaceNews.com, September 29, 2017, https://space
news.com/musk-unveils-revised-version-of-giant-interplanetary-
launch-system/.

17 "Mars," SpaceX, September 20, 2016, https://www.spacex.com/
mars.

18 Caroline Haskins, "The Legal Battle to Colonize Mars," The
Outline, March 15, 2018, https://theoutline.com/post/3739/mars-
colony-settlement-spacex-elon-musk-trump?zd=1&zi=343rx3r4.

19 Author interview with Charles Cockell, professor of astrobiology, University of Edinburgh, December 21, 2018.

20 Ray Bradbury, *The Martian Chronicles* (London: Folio Society,
2015).

Index